THE
Lady
AND HER
Monsters

THE
Lady
AND HER
Monsters

A Tale of Dissections, Real-Life Dr. Frankensteins, and
the Creation of Mary Shelley's Masterpiece

ROSEANNE MONTILLO

wm

WILLIAM MORROW
An Imprint of HarperCollins*Publishers*

HarperCollins books may be purchased for educational, business, or sales promotional use. For information please write: Special Markets Department, HarperCollins Publishers, 10 East 53rd Street, New York, NY 10022.

FIRST EDITION

Designed by Jamie Lynn Kerner

Art throughout courtesy of the Wellcome Library, London.

Library of Congress Cataloging-in-Publication Data

Montillo, Roseanne.
  The lady and her monsters : a tale of dissections, real-life Dr. Frankensteins, and the creation of Mary Shelley's masterpiece / Roseanne Montillo. — 1st ed.
    p. cm.
  Includes bibliographical references.
  ISBN 978-0-06-202581-4 1. Shelley, Mary Wollstonecraft, 1797–1851—Friends and associates. 2. Shelley, Mary Wollstonecraft, 1797–1851. Frankenstein. 3. Women and literature—England—History—19th century. I. Title.
PR5398.M57 2013
823'.7—dc23

                                                      2012021509

    13 14 15 16 17   ov/rrd   10 9 8 7 6 5 4 3 2 1

For my mother, Celeste Montillo; my late father, Giovanni Montillo; and my sister, Francesca Montillo

Contents

THE
Lady
AND HER
Monsters

PROLOGUE

*For I am building in the human understanding a true
model of the world, such as it is in fact, not such as a
man's own reason would have it to be; a thing which
cannot be done without a very diligent dissection and
anatomy of the world.*

FRANCIS BACON,
THE WORKS OF SIR FRANCIS BACON

CAMILLO'S FOOTSTEPS ECHOED LOUDLY AS HE CROSSED the empty cobblestone streets of Bologna toward his uncle's house. The afternoon was hot, and the scorching heat, coupled with that lazy midafternoon spell between noon and evening, allowed him to go by virtually unnoticed.

Summer in the city often proved vicious, and Bologna, the capital city of the northern Italian region of Emilia-Romagna, was suffering beneath the punishing sun. Harsh sunlight fell onto the red rooftops, distressing those inhabitants who had sought relief beneath one of the city's many porticoes. The city had been built on the edge of Via Emilia, a military road constructed by the Romans in 187 B.C. This extended from Piacenza, at the northernmost end, all the way to the Adriatic Sea. To the south, the city was bracketed by the Apennines, mountains that were as capricious as they were majestic, and to the north by the fertile lands and deep planes of the Po Valley, used for agriculture, farming, and raising livestock, particularly pigs. This had allowed Bologna to garner one of its many nicknames: "Bologna, la Grassa," "Bologna the Fat."

But as Camillo made his way across Piazza Maggiore, his mind was not on pigs; rather, it was on frogs.

He had been eagerly waiting for this day, but this infernal heat had been going on for weeks. Earlier, when he poked his head out of a window, something unusual had occurred: he heard the boom of thunder as it broke somewhere in the lowlands, and recognized the rumbles as they slowly made their way toward the city. Looking upward, he had noticed that thick dark clouds were slowly covering the sky. He then left his house and headed for the home of his uncle, the famed Luigi Galvani, one of the most renowned physiologists and obstetricians in Bologna.

Galvani had also been waiting for storms. As of late, he had been delving into experiments that complemented the medical and surgical skills he practiced in the city's many hospitals. But for almost a decade now, he had also been studying the field of *elettricità animale*, animal electricity. As a doctor, the field of electricity in general interested him deeply, particularly as it related to the cure of paralysis.

This so-called cure had been shown to work before. The Bolognese physiologist Giuseppe Veratti had applied electricity to various diseases, including paralysis and arthritis. The positive results were then set down in a book published in 1748 that Galvani had most certainly read.

Late in the 1760s, Veratti's experiments had also included the use of frogs and other small animals. Although those particular findings were never published, he gave demonstrations at the famed Academy of Sciences in Bologna in the years 1769 and 1770, where, at the time, Galvani was a member of and professor at the Institute of Sciences.

Galvani's choice of experimental animals also included frogs, what Hermann Helmholtz in 1845 called "the martyrs of science." Thousands of these amphibians had been slaughtered in the company of Galvani's nephews, assistants, and wife, and eventually he arrived at one conclusion: there was a fluid inherent to all living creatures that ran from head to toe, and this could be manipulated with an outside apparatus, such as a metallic arc or a rod. This

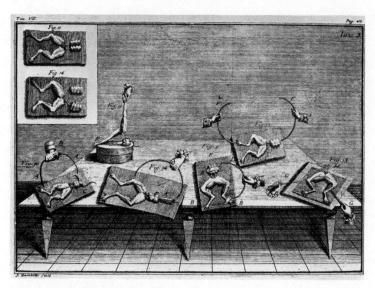

Lithograph from *De virubus electricitatis in motu musculari commentarius,* displaying the dissected frogs Luigi Galvani used in his experiments as well as the metallic arcs.

manipulation allowed the body to restore its inner activity, which in turn aided in the cure of paralysis and other diseases, restoring vitality.

In time Galvani found that pharmacology also influenced the results. In a lecture he delivered at the Academy of Sciences, he spoke of the effects opiates had on animal electricity. According to his notes, he injected opium into the frogs' abdominal cavities, stomachs, or cerebrums. While at first the frogs remained splayed and flaccid, they eventually revived and demonstrated a "violent convulsion, either from a slight tremor of the surface upon which they were resting or from contact with some body." If he hacked off the frogs' heads and pumped their bodies with opium, he got the same results.

But on that hot day, August 17, 1786, he wanted to prove a dif-

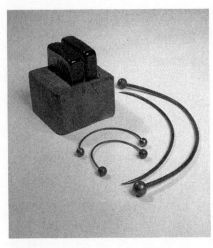

Apparatus formerly used by Luigi Galvani. Brass discharging arcs used to connect muscle and nerves.

ferent theory, using a method that in part resembled Benjamin Franklin's famous kite experiment: he wanted to see if he could elicit movements in the frogs' legs by employing atmospheric phenomena.

Galvani, his wife, Lucia, and his two nephews, who also served as assistants, gathered atop the balcony, the highest point in his house. Galvani had ordered metal hooks hung on the iron railings, and now a "prepared frog" dangled from each. The prepared frogs, he said, "should be cut transversally below their upper limbs, skinned and disemboweled . . . only their lower limbs are left joined together, containing just their long crural nerves. These are either left loose or free, or attached to the spinal cord, which is either left intact in its vertebral canal or carefully extracted from it and partly or wholly separated."

He gave the frogs the same close attention he gave all of his patients, employing the same expert surgical skills he had obtained at the hospital of Santa Maria Della Morte, "Saint Mary of Death," and Sant'Orsola. The movements of his hands were fluid, sinuous, and virtually flawless.

Notebook in hand, Galvani took copious notes. Those who knew him were aware that he brought an almost religious fervor to his work. Of course, that was not a coincidence. In his childhood, he had wanted to devote his life to God, a life of obedience and order, going so far as to join the Oratorio dei Padri Filippini, a religious order. But, as the firstborn in the family, his father, Domenico Galvani, and his mother, Barbara, who followed protocol and sent him to university, had chosen his path. It was at the University of Bologna, in the Faculty of Arts, that he came to realize the possibility of finding spiritual solace in scientific work. The university had been founded around the end of the eleventh century and eventually became known all over the world as "Alma Mater Studiorum" for being the oldest university in the western world. It boasted scholars and researchers in many fields, from law to philosophy, but the study of medicine, particularly anatomy, eventually made it notorious.

Lucia was the daughter of one of the most famous anatomists in the city, Domenico Gusmano Galeazzi. Unlike most children, Lucia had grown up privy to her father's experiments, which often involved mangled corpses being anatomized in a laboratory close to the family kitchen. During her youth, city officials had not precisely condoned anatomizations in private homes, though they had not discouraged them either.

They had pretended not to notice when the weather warmed that the rotten stench of decaying bodies trickled out, the sickening odor of putrid flesh mingling with the cooking smells of the city. Even then, officials assumed that anatomists were learning information that would be useful not only for their students, but for people at large. They were correct. Through one of those experiments, Galeazzi first detected the presence of iron in the blood and made even more discoveries regarding the body's gastrointestinal system.

As the boom of thunder neared, Galvani and his crew noticed the amphibians' legs twitching, and as he later reported in his *Commentaries,* "Just as the splendor and flash of the lightning are wont, so the muscular motions and contractions of those animals preceded the thunders, and, as it were, warned of them."

The frogs behaved as expected, and "in correspondence of four thunders, contractions not small occurred in all muscles of the limbs, and, as a consequence, not small hops and movements of the limbs. These occurred just at the moment of the lightning."

Although the frogs were dead, skinned, and nearly eviscerated, when zapped by an electrical arc or when they came in contact with a distant flash of thunder, their legs twitched in a way that made them seem as if they were ready to hop off the balcony and into the streets below. While this was happening, Galvani's tempestuous nephew Giovanni Aldini looked on. Standing on that balcony, feeling the hot August wind, watching black clouds roll above his head, and hearing for the first time in days the slight pinging of rain on Bologna's rooftops, he must have realized that these experiments were more than just his uncle's folly: they were the very essence of life.

In those moments, the theories that would direct the rest of his life began to form: Could those frogs truly return to life? And if that happened, what were the implications? Could his uncle's ideas later be used on lambs, oxen, sheep, and cows? And even further, could men benefit from such a thing? Not only the living, but also the dead? Could the process of reanimation be proven possible?

In the years that followed, Giovanni Aldini further tested those theories. The climax of his experiments occurred on January 17, 1803, at the Royal College of Surgeons in London, where he performed a never-before-tried experiment on the body of a convicted felon. By then, many of his earlier experiments performed on animals and humans—some dead and some living—had convinced him that galvanism (the new science named after his uncle) presented an opportunity for restarting one of the body's main vital organs: the heart. If that were to happen, the dead could reawaken.

By then, Aldini's experiments, and the topic of reanimation in general, had become fashionable in all of European society, from the natural philosophers, who began to delve deeper into the powers and possibilities of a vital force existing in humans and nature, to the more amateurish individuals, whose dubious endeavors merely allowed for a massive slaughter of frogs, pigs, and dogs, all in the name of science. It also became a go-to subject not only in the scientific community, but also among artists and writers and at crowd-pleasing soirées and salons all over England, France, and Germany.

In the early 1800s, the distinction between a scientist, an artist, a political reformer, and a man of letters was not as clear cut as it later became. The disciplines intertwined, the interests overlapped. As such, scientists like Humphry Davy and Erasmus Darwin not only studied the topics of electricity and vitalism, but also wrote poems and essays on the subjects, which were published and well received by the public at large. Poets such as Percy Shelley experimented with galvanic electricity, poisons, and gases, later jotting down long poems and odes that mused on the sublime mysteries of the natural world and the awesome powers of lightning and thunder.

But one particular author, Mary Godwin Shelley, truly combined the urgency of scientific endeavors in the late eighteenth and early nineteenth centuries, the lure of forbidden knowledge, and the power of literary interpretation in her masterpiece, *Frankenstein; or, The Modern Prometheus.* That it was published in 1818 and written barely a year and a half earlier was not a coincidence. She was well aware of the scientific procedures occurring around her. She had heard of Giovanni Aldini's experiments (first from her father's friend the medic Anthony Carlisle, who took great interest in such events and is believed to have attended Aldini's experiments, and later from her lover, Percy Shelley) and of his uncle Luigi Galvani's theories on animal electricity.

She knew of Humphry Davy and was aware of his lectures and writings, even using them, and him, for inspiration in her work. She had also read Erasmus Darwin's early theories of evolution. More important, her lover, who later became her husband, Percy Shelley, a poet, science aficionado, and fan of the macabre, was the one who introduced her to many of the scientific properties and theories exploding around her. He even went so far as to demonstrate certain experiments to her. Along with all of that, the literary publications of the time provided her with a good foundation.

Thus, it is no surprise, given all Mary Shelley had at her disposal, that she was able to create the archetype of the famed, mad, brilliant scientist of the nineteenth century: Victor Frankenstein.

Chapter 1

THE SPARK OF LIFE

Her lips were red, her looks were free,
Her locks were yellow as gold;
Her skin was as white as leprosy,
The Night-mare Life-in-Death was she,
Who thicks man's blood with cold.

SAMUEL TAYLOR COLERIDGE,
THE RIME OF THE ANCIENT MARINER

ON SUNDAY, AUGUST 24, 1806, MARY GODWIN AND HER younger stepsister, Jane Clairmont (later Claire Clairmont), hid quietly beneath the couches in the parlor. Their home, a five-story brownstone located on Skinner Street, had quickly become a stimulating hub for intellectual discourse, with Mary's father, William Godwin, the celebrated writer and reformer, as its master of ceremonies. The girls had not been invited to join the festivities, and no one knew they were even present in the parlor as a great bustle took place about them. Having snuck in there when no one was looking, they worked hard not to be discovered.

Jane Clairmont—Jane's mother and Mary's new stepmother—had forbidden them from attending these gatherings. Jane believed the conversations that took place among the crowd, on religion, the existence of God, politics, and the so-called principle of life, were inappropriate for young ears. Given her propensity for arguments, the girls often did as told, though sometimes Mary disregarded her stepmother and listened to her father's discussions from atop the staircase.

But on that Sunday, the poet Samuel Taylor Coleridge had arrived on Skinner Street, and Mary knew he would be reciting verses from his famous poem *The Rime of the Ancient Mariner*, published some years earlier in 1798. She had heard her father speak of it and now wanted to hear it for herself. Learning of her plan, her stepsister naturally followed suit.

Coleridge had met William Godwin in 1794, but like many

who crossed paths with the reformer then, he had not been impressed. "He appears to me to possess neither the strength of intellect that discovers truth, or the power of imagination that decorate falsehood," Coleridge had said. "He talked futile sophism." But after meeting Godwin again following the death of Godwin's first wife, Mary Wollstonecraft, Coleridge had changed his mind. Apparently, her death had mollified Godwin's character, softening his dark edges and making him more tolerable.

As the two girls eavesdropped, Godwin and the rest of the gentlemen gathered in. The wood-paneled room had been filled with a great deal of brilliance before: Humphry Davy, William Wordsworth, Charles Lamb, and many others. Now Coleridge took his turn. The poem was a mixture of poetical and popular language that some critics argued had come about as a direct response to the German gothic horror tales that were now so popular in England. *Rime* had appeared some years after the publication of Gottfried Bürger's *Lenore*, which had awakened popular fascination with the macabre. Writing some years later, a reviewer in the *Monthly Review* tied the publication of the *Rime* to "a time when . . . 'Hell Made Holiday' and 'Row heads and bloody bones' were the only fashionable entertainment for men and women."

Others speculated that the poem had been inspired by the explorations of James Cook, particularly his second voyage into the South Seas. This notion was further bolstered because William Wales, the astronomer on Cook's ship, was also Coleridge's tutor. Perhaps Wales had told Coleridge about those experiences. But others argued that Coleridge, who was often plunged into the depths of great depressions and anxious fits, had found his muse in the massive amounts of opium he had used to relieve these symptoms and that might have worked as a kind of hallucinogenic.

The girls huddled closer together as Coleridge began his story of an old ancient mariner who at first had been eager to leave his home in search of new continents. Upon his return from those explorations, he became adamant about telling his tale. While out walking one day, he stopped a man on his way to a wedding and recounted his travails. The man was indulgent for a time and found himself avidly listening and experiencing all the emotions a person went through in a lifetime: he was at times exhilarated, envious, and fascinated, and at others he felt sadness, sorrow, and even anger. And it is possible Mary and Jane, beneath the sofa, also suffered the same shifts in emotions as Coleridge's voice rang out:

The Wedding-guest sat on a stone:
He cannot choose but hear;
And thus spake on the ancient man,
The bright-eyed Mariner.

Coleridge captivated his listeners as the mariner recounted excitedly what good luck his ship had initially encountered. But it wasn't long before a mighty storm arose in the seas and blew the ship off course, driving it southward toward Antarctica, and the startled crew found relief in the sudden appearance of an albatross, which mysteriously began to guide them away from the bleak land of ice. Naturally, the sailors soon began to think the albatross had brought them much-needed good luck. But, watching from afar, the mariner became disgusted by his crew's show of superstition.

Angry, he lifted his eyes toward the albatross and, in a moment of unbridled passion, shot it dead. The crew became distressed and began to wail in despair, and as if echoing their own agony, the spirits swirling around them began to grieve the great abomination that had been committed against nature. To further inflict punishment, the spirits followed the ship through unfamiliar waters.

The poem moved ahead as the vessel did, Coleridge's voice most likely rising and falling as the waves continued to lull the ship; it was then that he described the encounter between the crew and the eerie vessel boarded by Death and the Night-mare Life-in-Death. A struggle ensued, and nature struck back, killing all but the mariner. Filled with guilt for killing the albatross and the consequences of that act, the mariner was left doomed to wander the earth forever, always repeating his tale as a final act of atonement:

> He went like one that hath been stunned,
> And is of sense forlorn:
> A sadder and a wiser man,
> He rose the morrow morn.

The girls stayed quiet as the tale ended and the grown-ups began debating the various properties revealed in the poem: the images of life and death; the mysteries of sin, redemption, and the repercussions of guilt; the pursuit of forgiveness and forbidden knowledge; the ache and sorrow that loneliness brings; the belief in superstitions, of atoning for one's sins. A decade later Mary Godwin would use similar imagery in the opening scenes of her most famous novel, *Frankenstein; or, The Modern Prometheus*. In it, the fictional character of Robert Walton, a mariner and explorer intent on finding a passage to the North Pole, appears and echoes Coleridge's mariner as he too was traveling into uncharted waters, trying to be the first explorer not only to accomplish such a feat, but to do so while avoiding a mutiny on the ship. Coleridge's mariner is mentioned again when Walton, in writing to his sister Margaret, declares, "I am going to unexplored regions, to 'the land of mist and snow,' but I shall kill no albatross, and therefore do not be alarmed for my safety."

But many others were to inspire Mary Shelley in the writing of *Frankenstein*, though there was no indication of them yet. That evening she only knew that *The Rime of the Ancient Mariner* had made a deep impression on her soul, one that would last a lifetime.

ON THE NIGHT OF MARY GODWIN'S BIRTH, AUGUST 30, 1797, A STORM descended upon the city of London that was later remembered as one of the most awesome displays of thunder and lightning anyone had ever seen. Loud, crackling noises pierced the night air, while jagged yellow lines crisscrossed the inky night sky. It was a wondrous spectacle Mother Nature seemed to revel in, and some were awed by it. The story of Benjamin Franklin's "stealing" thunder from the sky in 1752 was widely known and still played havoc in people's imagination, allowing them to believe in the "testimony to the ability of human reason to bring nature under its sway." Natural philosophers like the famed Humphry Davy also saw it as a vehicle not only to understand nature, but to "interrogate [her] . . . not simply as a scholar, passive and seeking to understand her operations, but rather as a master, active with his own instruments."

But others, given their superstitious and religious mind-set, were frightened by nature's so-called wonders. To them, the idea that nature could be made to bow down to man bordered on the sacrilegious. If man could steal thunder from the sky; elicit electricity from the heavens; make dead frogs, sheep, and dogs jump; and impart a certain measure of respiration to the dead, then what need was there for a God who had dominion over everything and everybody? These people believed the angry thunderstorms of August 30 were a sign not of untamed knowledge, not of nature bending down to human will, but of God's wrath. The human race had overstepped its boundaries in some fashion, and God was now seeking His vengeance.

But in the Godwin home in the Somers Town district of London, neither idea was truly being contemplated. Those living within it thought of the powerful show outside their window as just a storm, a vicious storm that coincidentally was occurring on the night of the baby's arrival.

Mary Wollstonecraft's labor pains had begun earlier that day, when she retired to her bedroom just before two P.M. Feeling a nagging ache in the lower portion of her back, she slowly hiked up the staircase, aware of what to expect. Having gone through a pregnancy and childbirth before—her firstborn daughter, Fanny, was now three—she knew what would happen in the hours ahead. No male doctor would be present at the birth. Instead, she had decided to have only a woman midwife to "sit by and wait for the operation of nature."

Her husband, William Godwin, waited downstairs. He had been a bachelor until the age of forty-one, and his rigid and somewhat inflexible manners had changed only upon the second meeting with, and subsequent awakening of his affection for, Mary Wollstonecraft. In his early forties he became a husband, a stepfather to Fanny, and a father-to-be. As they settled into a life together, Godwin tried to find his way among his new roles, though he still had a certain measure of inadequacy about him. When Mary's labor started, he was happy to remain below.

As the afternoon and evening progressed, the storm intensified, much as Mary's labor did. Both, it seemed, were gathering momentum, and at about nine o'clock Mrs. Blenkinsop arrived to serve as midwife.

Mary's labor progressed relatively normally, and at twenty past eleven a baby girl was born. As the etiquette of the time required, William Godwin was asked not to enter his wife's room until all stages of labor and delivery were over. He waited patiently but anxiously watched the hours slowly ticking away, night ebbing into dawn. Eventually he heard the midwife's footsteps rushing toward him; she told him "that the placenta was not yet removed." Unable to continue on her own, she advised Godwin to search for another doctor, this time a male one.

The storm outside continued to rage on as a frightened Godwin rushed to call on Dr. Poignand, who arrived at the house several hours after the baby was born.

The eighteenth century was a remarkably difficult time for mothers and their infants. Infections, mistakes, malnutrition, and lack of care before, during, and after a pregnancy all resulted in a surprisingly high number of deaths. Dr. Poignand was a typical physician of the era and did what he could under the circumstances. Arriving in Mary Wollstonecraft's chamber, he made a few disparaging comments about delivering a child without the aid of a male physician nearby. Then he rolled up his sleeves, raised the dampened sheets that covered Mary Wollstonecraft's sore body, and, without latex gloves, inserted a hand between her naked legs. Slowly, the doctor removed Mary's placenta piece by bloody piece, pushing his dirty hand several times within her vagina.

He then became convinced that he had removed the entire placenta and assured William Godwin that everything would be okay. Writing afterward, Godwin recalled "the period from the birth of the child till about eight o'clock the next morning . . . full of peril and alarm. The loss of blood was considerable, and produced an almost uninterrupted series of fainting fits."

Dr. Poignand had been incorrect when he said he removed all of the placenta; a chunk had been left behind in Mary's womb and was now festering. A new doctor, Dr. Fordyce, arrived later and said Mary's condition was so grave it was not safe for her to nurse the new baby. Puppies had to be brought in to "draw out the milk" from her swollen and painful breasts.

For the next several days, she lingered between this world and the next. At certain times William Godwin felt hopeful, but during Mary's shivering fits, despair overwhelmed him, and he knew "every hope was extinct." At one point, he asked Mary what direction "she might wish to have followed after her decease."

What did she wish for her two small daughters? William approached the subject carefully, proclaiming that she was very ill and would take a considerable time to recover. But Mary knew what he was asking.

"I know what you are thinking of," she replied with little strength. She did not go any farther.

On September 10, at 7:40 A.M., Mary Wollstonecraft, the first and most influential feminist and the author of *A Vindication of the Rights of Woman*, died of puerperal fever at the age of thirty-eight, in the same bed where eleven days earlier she had given birth to her daughter Mary Godwin, later to become Mary Shelley. Mary Wollstonecraft was buried on September 15, 1797, in the old St. Pancras churchyard.

William Godwin did not attend the ceremony. Bereaved and full of "longing," he tried to get his mind off the alarming thoughts that had overwhelmed him since Mary's death. One of his wife's books was near him, but he did not want to pick it up, much less read it. Instead, he focused on another book by his side, this one detailing "the education of children." As he leafed through it, he could not help but think of the "two poor animals" who were now his sole responsibility.

Rather than reading, he decided to write a letter to the medical man Anthony Carlisle, a friend of his and Mary's. "It is pleasing to be loved by those we feel ourselves impelled to love," Godwin wrote. "It is inexplicably gratifying, when we find those qualities that most call forth our affections, to be regarded by that person with some degree of feeling."

THAT TWO SUCH INDIVIDUALS—MARY WOLLSTONECRAFT AND William Godwin—became entangled with one another to begin with struck some as an utterly peculiar event. They had initially become acquainted at a dinner party that Godwin attended to meet Thomas Paine, who had just published *The Rights of Man*. It was not love at first glance for William and Mary. On the contrary, they were "mutually displeased with each other." Godwin had hoped to spend the evening with Paine, and Mary's presence there irked him.

Mary was an attractive woman who was rather tall and had brown hair and eyes. But right away, Godwin was put off by a streak of gloominess that was part of her persona. She would pass this trait on to her daughters. She was left cold by Godwin's habit of complimenting everyone he met, even when they did not merit it. This was certainly not the most auspicious start of a love match in history.

They saw very little of each other after the dinner party, as Mary went to France to attend to some business. It was a personal matter about a man with whom she'd become infatuated: Henry Fuseli, a painter eighteen years her senior. She didn't seem to be bothered by the fact that he was married, and those around her did not understand why she was fascinated with him. This included Godwin, who thought Fuseli was not an intellectual but a snob.

In the winter of 1792, Mary decided the only way to have a deeper relationship with Fuseli was to include his wife, Sophia. She propositioned them with a sort of ménage-a-trois that would involve all of them living together and her becoming their mutual partner. Not surprisingly, they rejected her.

Toward the end of 1792, she was living a lonely existence in a tiny Paris apartment, the icy landscape of the city matching her own sadness. The passion she had desired from Fuseli may not have materialized, but she was desperate for the affection of any man. That's when she fell for the American Gilbert Imlay, who picked up on her vulnerability and need, which let him feed her mind with fantastic (and false) stories of his past and those of a future they might have together. She quickly fell for him and clung tightly to him, especially when she learned she was pregnant.

The pregnancy brought about a dramatic shift in their relationship, causing Imlay to spend weeks away from Paris, most especially in Le Havre. As the days turned into weeks and weeks extended into months, Mary's familiar ache and loneliness returned. Only toward the end of her pregnancy did Mary join Imlay in Le Havre, where her daughter Frances—Fanny—was born. In September

1795, soon after the birth, Mary left for London, in what she believed would become a permanent separation.

Not long after, Mary learned that Imlay had found another woman. She urged him to change his ways and meet their new baby daughter, but this did not happen. Again she was alone, but this time with a baby. Unable to continue on, Mary decided to end her life. "I have been treated with unkindness, and even cruelty by the person . . . [from] whom I had every reason to expect affection," she wrote to Archibald Hamilton Rowan. "I looked for something like happiness in the discharge of my relative duties, and the heart on which I leaned . . . pierced mine . . . I live but for my child, for I am weary of myself . . . I have been very ill—have taken some desperate steps . . . for now there is nothing good in store,—my heart is broken!"

During the eighteenth century the river Thames had become a major center of commerce by transporting goods across the Brit-

Panorama of the river Thames and the buildings of the city. In the eighteenth century, the river provided a great divide between social classes in London. It was also from one of its bridges that Mary Wollstonecraft jumped trying to commit suicide.

ish Empire and servicing farmers, fishermen, tradesmen, and other commercial ventures. It had also formed an unspoken boundary between the different classes who lived on either side of its waters. And that river, Mary decided, would finally transport *her* to her next life. It would become her grave.

She tried to find a quiet spot for her final moments but could not find one on the Battersea Bridge. The evening of her demise was a viciously cold and rainy one in October, a dreary occasion even by London's standards. Rather than being deterred, she decided this weather was helpful. Drenched, undoubtedly lonely, and surely frightened, Mary walked up and down the wooden bridge, allowing the rain to soak her clothes.

On this night, no one was on the bridge, which meant she could carry out her plans in secret. The rain that seeped through her clothes added much-needed weight to her frame. When she thought she was heavy enough, she neared the parapet. She felt the cold dark currents sloshing against the riverbank below were beckoning her, and she jumped. One would imagine her body, now soaked, would have sunk deeply and quickly, but that's not what happened. Agitated, she struggled against the currents and became tangled in her clothing more and more tightly until she passed out.

Her body washed ashore and was later found and revived by a passerby. Gilbert Imlay rushed to her, declaring his love, but strangely enough, Mary was not moved by this. Apparently, plunging into the cold water had shaken her out of her melancholy, and she realized the affair needed to come to some sort of resolution.

Around this time, she reconnected with William Godwin. Having been invited to take tea with Thomas Holcroft, she was surprised to see Godwin there as well. As before, their exchanges didn't cause either one of them to feel any flurry of love or passion toward the other. By now Godwin had become famous, which seemed to have boosted his demeanor. He was socially awkward but also bent on achieving fame and acceptance from society, so this new lifestyle provided a bonus.

On the other hand, Mary Wollstonecraft was now a disreputable woman with a sordid love life and an illegitimate child. Not surprisingly, Godwin didn't think she was as irksome anymore, but rather, somehow, the suffering Godwin saw on her pale features gave her an alluring, vulnerable quality, so much so that he was drawn to a sense of "sympathy in her anguish." In the following weeks, they saw a great deal of one another and eventually both spoke of "the sentiment, which trembled upon the tongue but from the lips of either."

To them, the state of their relationship felt as good as a marriage, without the restrictions of an actual ceremony. They both detested such shows of formality. "Nothing can be so ridiculous upon the face of it, or so contrary to the genuine march of sentiment, as to require the overflowing of the soul to wait upon a ceremony," Godwin declared. That is, until a child entered the picture.

When Mary became aware of her second pregnancy, she recalled the scorn she had suffered during her first. Godwin, of course, agreed to marry her, though doing so went against all the principles he had been advocating for years. He was aware that some would see him as a hypocrite for yielding to the institution he so despised: "Some people have formed an inconsistency between my practicing this instance & my doctrine," he wrote to his friend Thomas Wedgwood. But he also explained why he did not see any inconsistency. He still believed marriage was wrong, and he had only married Mary because he cared for her. Despite having gone through the ceremony, he felt no different than before and said, "I hold myself no otherwise bound than I was before the ceremony took place."

On March 31, 1797, William Godwin and Mary Wollstonecraft married in the small church at St. Pancras, and on April 6, 1797, they moved into a house together located at 29 Polygon Road, in London's Somers Town district. Godwin immediately began to receive congratulatory notes that said such a union was powerful and intellectually a fabulous match. One note came from Thomas Holcroft, in whose house they had reacquainted themselves. As others before him had, Holcroft extended a happy note to Godwin for having landed "Mrs. W." But whether or not his wishes were heartfelt remain unclear, because earlier in the months preceding the marriage, Holcroft seemed to have a great and passionate crush on Mary Wollstonecraft.

"I think I discover[ed] the very being for whom my soul has for years been languishing," he wrote to her. "The woman of reason all day . . . in the evenings becomes the playful and passionate child of love . . . one in whose arms I should encounter . . . soft eyes and ecstatic exulting and yielding known only to beings that seem purely ethereal: beings that breathe and imbue but souls: You are this being."

EVEN THOUGH WILLIAM GODWIN HAD AT FIRST BEEN RELUCTANT TO marry her, the void he felt upon Mary's death was deep and long lasting. And he dealt with it in the only way he knew how. Stunned into disbelief, the day after her death William Godwin entered his study, sat down at his writing desk, set quill to paper, and began working on what eventually became *Memoirs of the Author of "A Vindi-*

cation of the Rights of Woman." He felt obliged "to give the public some account of the life of a person of eminent merit deceased" during that particular time, for "it is a duty incumbent on the survivors."

His intentions were to tell Mary's story, to highlight her heartbreaks and successes, her triumphs and apologies, so that readers could glimpse the most transfixing woman he had ever known. He was well intended, though when the memoirs were published in January 1798, the criticism and backlash he received came as a stab in the heart. On those pages he had poured out his heart as well as Mary's secrets, going to great lengths to highlight not only her life but also her private affairs and indiscretions, including her infatuation with Henry Fuseli, her affair with Gilbert Imlay, and the birth of Fanny.

Most of the information had been taken from private letters, journals, and confidences shared during their conjugal life. He also included detailed accounts of her two suicide attempts and her bouts with depression, clearly something her readers, those who had read *A Vindication of the Rights of Woman,* had either not known about or wished not to know about. In Godwin's hands, Mary Wollstonecraft came across as a bit of a hypocrite: in her work, she had fought for the equal rights of women, for owning one's own life and doing with it what one may, for refuting marriage, for being on par with men, for having other choices; most of all, she had attacked the educational system of the time for training young women solely to be "the toy of man, his rattle, and it must jingle in his ears, dismissing reason, whenever he chooses to be amused." And yet in Godwin's book, she was attempting to drown herself over the inconsequential Gilbert Imlay? And what was to become of the child she had given birth to?

Written and published reviews were even harsher than the ones he received in person. One particularly nasty one printed in the *Monthly Review* declared that "blushes would suffuse the cheeks of most husbands, if they were *forced* to relate these anecdotes of their wives which Mr. Godwin voluntarily proclaims to the world. The extreme eccentricity of Mr. G.'s sentiments will account for his conduct. Virtue and vice are weighed by him in a balance of his own. He neither looks to marriage with respect, nor to suicide with horror."

Godwin also foolishly revealed that he and Mary had engaged in sex before their marriage. This particularly angered those whose strict religious beliefs went against the notion of premarital sex, so much so that the memoirs were dubbed merely "a narrative of his licentious amours."

Only years later did Godwin actually refer to the events that transpired upon the death of Mary Wollstonecraft as "stained with . . . melancholy colours . . . the air appeared to be murky and thick, an athmosphere that bore pestilence on its wings." By then the damage had been done.

Godwin's book misrepresented everything Wollstonecraft had worked so hard to accomplish, damaging her reputation and causing her works to be disregarded for years to come. Only her daughter Mary claimed kinship with her. She read her mother's works and her private letters in the seclusion of St. Pancras Cemetery and her father's study, and through them learned how influential her mother had been to other women. This influence would extend over Mary's own life, particularly when it concerned romantic matters. She would become bonded with her mother as she suffered the scorn and misjudgments of a society that would not understand her affair with Percy Shelley.

ᵔᵛᵎ

ON A WARM TUESDAY AFTERNOON IN MAY 1801, WILLIAM GODWIN
found himself sitting beneath one of his balconies reading a stack
of notes, when he heard a woman's voice from a nearby window
directed toward him. He looked up and saw a woman who was
neither young nor beautiful in the conventional sort of way, but
homely by most people's standards.

She must have smiled as she craned her neck forward and said,
"Is it possible that I behold the immortal Godwin?" Conceited and
easily duped by flattery, Godwin married the neighbor not long
after the initial encounter, on December 21, 1801.

William Godwin's desire to remarry had been kindled soon
after Mary's death. Though he had been content in his younger
days to study and write, he had enjoyed his late-in-life marriage.
And now he knew he could not go back to the way things used to
be: he was a lonely man in his forties, with two young daughters,
one of whom wasn't even his natural daughter, though he had given
her his name. Servants were aiding him in rearing the two "unfortunate little wretches," but it didn't take long to see he didn't possess the fortitude or sensitivity to be a single parent.

The job called for a special kind of woman. He met a few
women who he believed were worthy candidates to join his household, but for some reason or another, his advances were always met
with disappointments. He also wanted to return to his writing and
studies with a more methodical schedule, something that full-time
parenting didn't allow him to do.

True, in 1799 he had managed to publish the gothic novel *St.
Leon*, a tale dealing with isolation, immortality, alchemy, the elixir

of life, and the consequences of forbidden knowledge, but through it all the topic of child development and rearing remained at the back of his mind. He desired a wife, someone he knew would do a better job with the girls than he did while he continued on with his work. As it happened, the next Mrs. Godwin would come from right next door.

Mrs. Jane Clairmont, the woman who so famously uttered the words that overwhelmed William Godwin's soul, had a spotty past herself. Quarrelsome and highly business-minded, she found it difficult to keep her anger in check. Few believed she had any talent for ingratiating herself in the lives of others; if anything, she appeared to work hard to alienate those Godwin knew from his past, as well as his daughter, Mary, who in time came to believe Jane Clairmont's main duty was to draw her father away from her.

Her observations were not entirely untrue. Later in life Mary Shelley, in writing to her friend Maria Gisborne, said, "Mrs. Godwin had discovered long before my excessive & romantic attachment to my father."

Mrs. Jane Clairmont had declared herself a widow for several years, and she had two children close in age to Godwin's. It's doubtful that the two children, Charles Goules Clairmont and Jane Clairmont, had the same father, though their mother never spoke of it. She had also been searching for a new mate for some time; spotting the famous Mr. Godwin beneath the balcony must have seemed like the moment she had been waiting for.

It did not take long for Godwin's friends and acquaintances to decide that they were not pleased by the new woman in his life. "The Professor is COURTING," his friend Charles Lamb said. "The Lady is a widow (a disgusting woman) and the Professor has grown quite juvenile. He bows his head when spoken to, and smiles without occasion . . . You never saw . . . anyone play Romeo so unnaturally."

Unlike his marriage to Mary Wollstonecraft, whom they had all found intellectually challenging and formidable, this woman came across as horrendous, at the very least. This was not a bond sealed by intellectual debates or an amicable companionship, they all felt. Instead, it seemed like a business deal between two floundering individuals who had now the opportunity to begin again.

This fresh start, and a new son, William Godwin, brought them to the home on Skinner Street. In theory, the older five-story structure should have allowed for more space, more privacy, and a good measure of solitude. This would also be a perfect spot to open the bookshop the Godwins had in mind, as the street was lined with bookshops and other storefront businesses.

Unfortunately, they soon discovered that the house was not as ideally located as it sounded. It was near the city's prisons and courthouses, and on hanging days, which were quite popular with many Londoners, the family heard not only the banging and hammering of the gallows going up, but also the crowds as they gathered as early as dawn for a chance to catch even a glimpse of the proceedings, their desire for a good spectacle seeming to contaminate all that surrounded them. This frightened Godwin because the people began to work themselves into a frenzy, and by the time the convict arrived, they thirsted for blood, often with deadly results.

The hangings were gruesome, but the public still looked forward to them as if they were attending a sporting event. They were so common in London's squares that the barrister Charles Phillips said, "We hanged for everything—for a shilling—for five shillings—for five pounds—for cattle—for coining—for forgery, even for witchcraft—for things that were and things that could not be." If the execution was of a particularly notorious criminal, mobs numbering in the tens of thousands gathered outside the prison—often the Old Bailey—and waited around until the convict had seen his last moment.

The throngs came from all walks of life: the well-to-do paid handsomely for a spot near the actual gallows (it was said that from there it was easier to view the final twitching of the body and hear the actual gurgle of the man as the air left his lungs). Everyone else stood where they could, jostling and bumping into one another or going so far as to hang from windows and balconies to get a better view. Young officers stood toe-to-toe with mothers-to-be or those clutching small children at their breasts. Fathers brought their young sons with them as an occasion for male bonding. Some attended as a family and wore their Sunday best, the mothers carrying wicker baskets full of goodies and the fathers holding their children up so they could see over everyone else's heads. If the criminal's sentence also included an anatomization, the people followed the dead criminal as he was hauled to the site of the procedure and stood there until the dissections were complete.

On most occasions people behaved, viewing the proceedings in an orderly, if loud, manner, their hilarity echoing the tolling of the bells nearby, while a few murmured the words of the bellman of St. Sepulchre's on such occasion, whose job it was to announce the pending ordeal: *"All you that in the condemned hold do lie / Prepare you, for tomorrow you shall die. / Watch all and pray, the hours is drawing near / That you before the Almighty must appear. / Examine yourselves, in time repent, / That you may not to eternal flames be sent / And when St. Sepulchre's bell tomorrow tolls, / The Lord above have mercy on your souls."*

They often jeered louder as the convicts were marched up to the scaffolds. But sometimes, when the masses were unusually large, matters turned rough, as happened on February 4, 1807. On that day, people came from all over the city and the surroundings to see the hangings of John Holloway and Owen Haggerty, who were found guilty of killing a man and sentenced to hang. The fatal testimony had come from one of their accomplices, who had ratted them out in exchange for his own life. Many thought the two men were innocent and were being railroaded by a failed judicial system. But still the crowds gathered for their execution, and "by eight o'clock, not an inch of the grass was unoccupied in view of the platform."

When the men arrived, the crowd pressed forward and a "terrible occurrence" took place. According to the *Newgate Calendar,* a monthly calendar that provided notices on executions, "the pressure of the crowd was such that, before the malefactors even appeared, numbers of persons were crying in vain to escape it . . . Several females of low stature who had been so impudent as to venture among the mobs were in a dismal situation; their cries were dismal. Some who could be no longer supported by the men were suffered to fall, and were trampled to death."

The chaos continued as men, women, and children were pushed, shoved, stomped upon, suffocated, and crushed to death, and "those who once fell were never more suffered to rise." The area was later cleared, after the two men were hung and left to dangle for an hour. When the stream of onlookers dispersed, "there lay nearly one hundred people dead, or in a state of insensibility strewn about the street." Later, it was determined that the people had died from "compression and suffocation."

Godwin did not attend these spectacles, but he did correspond with some of the prisoners and even considered visiting one of them. Around the mid-1790s, he wrote to an unnamed prisoner in the Tower of London whom he wanted to possibly see. After some squabbles with those in charge, Godwin reconsidered the visit but kept up his correspondence. The letters themselves seemed to be an effort made to lift the prisoner's mood, but to a man shackled in the fetid cell, the appeal probably came across as an unsympathetic spew of doctrine. For instance, Godwin told the prisoner he hoped this time alone would allow him to "reflect on his error." Perhaps being imprisoned would let him feel "the beauty of universal benevolence." He seemed to lack an understanding of why the prisoner felt so embittered and angry.

Though the recipient of Godwin's letters remains unknown, there are two likely candidates who might have stirred his interest at the time. One was the political prisoner John Augustus Bonney. While a prisoner he was also immersed in writing a history of the Tower and all that occurred to him while he was there. At one point, he was transferred to Newgate Prison, though he later returned and continued his writing. But there was also a reverend, John Tooke, who was accused of having been a member of the London Corresponding Society, a political society. Tooke had also been imprisoned in the Tower. He also kept a journal and jotted down a history of the Tower.

Though Godwin wrote to one of these men, he never went to any of the hangings. Instead he secluded himself in his study, and on Sunday afternoons, he held intellectual gatherings in his parlor where his friends, such as the medic Anthony Carlisle, as well as the up-and-coming young natural philosophers of the day, discussed works and ideas that were stirring the public's mind and imagination. One person whose works were being discussed was Humphry Davy.

GODWIN MET DAVY IN EARLY DECEMBER 1799. THEY WERE INTRO-duced by Samuel Taylor Coleridge, who had befriended the young scientist not long before. Coleridge had become so enthralled with Davy, he had begun to delve deeper into the mysteries of chemistry. But Coleridge was as interested in learning about chemistry as he was in becoming even closer to Davy.

And this bond had developed because of the poet's desire to get high. In the late 1790s, Humphry Davy had been hired by Thomas Beddoes to work in the newly established Pneumatic Institution in Bristol. Beddoes, a notorious physician who made a living catering to the wealthy of that city, had set up the institution as a place to study the various properties of gases and how their administration on the body could be used to cure various diseases. As it happened, he was in need of an assistant to mind the well-stocked laboratory. Earlier that year, Humphry Davy had become friends with Gregory Watt, son of the famous engineer James Watt, as well as Davies Giddy, who was friendly with Thomas Beddoes. They introduced Davy to Beddoes.

At the Pneumatic Institution, the notion that gases could some-how be linked to the vital powers took root, and Humphry Davy was asked to perform experiments to prove or disprove that link.

Davy realized this was a great opportunity. "Who," he later wrote, "would not be ambitious of becoming acquainted with the most profound secrets of nature; of ascertaining her hidden operations; and of exhibiting to men that system of knowledge which relates so intimately to their own physical and moral constitution?"

On April 9, 1799, Humphry Davy inhaled sixteen quarts of

nitrous oxide gas, commonly referred to as laughing gas. Predict-
ably, the gas worked as a hallucinogenic, and he was so surprised
and delighted by the results that he "resolved to breathe the gas for
such a time, and in such quantities, as to produce excitement equal
in duration and superior in intensity to that accomplished by high
intoxication from opium or alcohol."

In the cocoon of his laboratory, the inhalation of the gas produced the desired effects. "I existed in a world of newly connected
and newly modified ideas. I theorized; I imagined that I made discoveries . . . ," Davy later wrote. "My emotions were enthusiastic
and sublime . . . As I recovered . . . I felt an inclination to com-
municate the discoveries I had made during the experiment. I endeavored to recall the ideas . . . *Nothing exists but thoughts! The universe is
composed of impressions, ideas, pleasures and pains.*"

As more and more people heard about the institute's and Davy's
experiments, the idea of indulging in such feelings of benignity
and mental stimulation appealed to a select group of individuals.
One of them was Samuel Taylor Coleridge, who rushed to Davy
straightaway.

Not long after the experiments, Davy left Bristol for the recently founded Royal Institution in London. There, his scientific
pursuits and dazzling displays of chemistry quickly earned him a
reputation as one of the scientific leaders in London, his lectures
always a well-received mixture of the practical and the spectacular.
Electricity featured prominently in his lectures, as did the possibility of reanimation.

IN 1802 DAVY DELIVERED A LECTURE AT THE ROYAL INSTITUTION
that was later published as "Discourse Introductory to a Course

of Lectures in Chemistry." In it, he described the latest findings about man's ability to conquer nature, and about the so-called vital powers.

Davy had argued that chemistry had "given to him an acquaintance with the different relations of the parts of the external world; and more than that, it [had] bestowed upon him powers which may be almost called creative; which have enabled him to modify and change beings surrounding him, and by his experiments to interrogate nature with power."

Davy believed this exploration could be accomplished not only through a deeper understanding of chemistry and the vital powers, but also by using the latest experiments, which included the quest for reanimation. Galvanism, it was called, a method of study and experimentation employed earlier by the famed Luigi Galvani.

Galvani, along with his assistants at the University of Bologna, had been able to imbue dead frogs with respiration and animation, accomplishing what others had thought impossible. Should they not endeavor to continue his work? Davy asked. And of course, along with him, others not only speculated but agreed.

Chapter 2

WAKING THE DEAD

*Lastly, where is that most noble
electrical fluid that seemed entrusted
with motion, sensations, blood, circulation,
in short, with life itself?*

LUIGI GALVANI, "CARNIVAL LESSONS"

DURING BOLOGNA'S CARNIVAL SEASON OF 1786, LUIGI Galvani stood in the anatomical theater, rhythmically sharpening his surgical tools. A large crowd had assembled to view the lectures and demonstrations, and they milled about in the wood-paneled room, while the wealthier members of the city had already seated themselves in the soft-cushioned chairs near the center of the auditorium. These public demonstrations were open to everyone, but the groups had been segregated according to social rank, with the rich sitting near the spot where the cadaver lay splayed in the circular center of the theater and the poorer classes viewing the proceedings from the far back.

It was a good place for them to be. Their presence, with its snickering and loud groans, enlivened this somber occasion and was clearly heard among the presentations. If the cadaver had been someone notorious, they felt lucky. If he happened

LUIGI GALVANI

Iconographic lithograph of Luigi Galvani at the height of his academic career.

to be someone they had known personally while alive, so much the better, for they could judge for themselves whether or not such a public ending was warranted. More often than not, they felt, the ending was appropriate.

The crime rate in Bologna during the late 1700s was especially high, due mostly to gangs of bandits that ruled the city and were oblivious to its laws. They crept down the swaths of darkness created by the city's kneeling buildings, or across its shady alleyways, and robbed people at knifepoint as they crossed Bologna's many squares and piazzas, especially the farmers, who left their homes as first light dawned and returned when darkness descended. Some gangs were more vicious than others, and they would commit murder for nothing more than a basket of vegetables, a dozen eggs, a dead rabbit.

They had the best hiding spots, places the law never thought to look, such as sacristies within the churches or the cells of monasteries. But if one was actually caught, he was then tried, convicted, and sentenced to death. The death would be very public, announced in bulletins plastered all over the city. These advertisements of "hanging, quartering and beating" not only scared would-be criminals but also invited the community to participate in the big event.

Punishments for minor crimes may have been less severe, but they were no less painful or public. The insertion of spikes (wooden or iron) between the fingernails and the skin, or the toenails and the skin, was a favorite form of torture until the early 1800s. Sometimes angry dogs were unleashed on criminals until the felons, bitten and pawed nearly to death, had no other choice but to confess. But before the punishments were inflicted, the felons were piled on horse-drawn carriages and paraded across the city where the citizens, commoners as well as those belonging to the nobility, had an opportunity to view them. Some were specifically hanged or beheaded, and those were the ones brought beneath the anatomist's knife.

The crowd in the anatomy theater waited anxiously as the anatomist prepared the tools of his trade. This loud show of anticipation brought some laughter and nervous coughs from the rear of the building, a reaction that those sitting up front did not appreciate. Those coveted seats were filled by the members of the *podestà*, the town's officials, the doctors, the students, the scholars, the fine ladies who donned their puffy-sleeved silk gowns with baubles on their fingers or wore fancy carnival masks. Outside, the bells from across the street tolled and the melodious strains of monks reciting a rosary sounded in the auditorium, in hope for the peaceful rest of the cadaver's soul.

Each year an official member of the anatomy team was chosen with great fanfare to demonstrate his skills to his peers, students, and the public at large. He gave fourteen oral lectures but also showed practical dissections of human corpses. It was no coincidence that they took place during carnival season—late January to mid-February. Carnival ushered in Lent, and by nature this was a time of sanctity and revelry, giving the demonstrations the double-edged feeling of the sacred and profane. But there were also more practical reasons: students were excused from their official studies then and could attend the lectures without guilt. But more to the point, the dead preferred the cold weather, which lessened the decomposition and smell of a dissected cadaver on a marble slab.

Luigi Galvani led these ceremonies on four occasions: 1768,

1772, 1780, and 1786. Cutting deeply into the flesh with long, secure, sinuous strokes, the presentations were proud moments that allowed him to share his thoughts on his latest experiments with his colleagues, who then had the opportunity to dispute his claims. As he spoke, he must have realized that these demonstrations were somewhat lurid, especially when the anatomists themselves gave vent to their own speculations and feelings.

Most students knew that in 1521, Professor Jacopo Berengario da Carpi had described the gruesome dissection of a woman, and then detailed it in a book called *Commentaria*. The story trickled down within the halls of the university and became notorious among students and professors. It told of Berengario's removal of a woman's placenta while performing a public demonstration, of how he had held it aloft "before almost five hundred students and our University of Bologna and also many citizens." The viewers were riveted as the professor brought the woman's entrails out of the rib cage toward the open air. He seemed to find nothing odd in what he was doing; to him, the anatomist was not only someone poking the flesh and prodding the innards of a corpse, "but a philosopher who investigates the secrets of nature."

As the crowd watched Galvani prepare to dissect the cadaver, one thing became quite obvious: the Bolognese were not a sheepish people. Even earlier than Berengario da Carpi's demonstrations, in the year 1315, medical students attended human dissections performed by Mondino De' Luzzi, whose first corpse had also been a woman. Not surprisingly, those procedures were later described in his book *Anatomia Corposi Humani*. An enthralled witness to Mondino's event, Guy de Chauliac, described the ordeal: "the body, having been placed on a table, he would make from it four readings: in the first the digestive organs were treated, because more prone to rapid decomposition; in the second, the organs of the respiration; in the third, the organs of the circulation; and in the fourth, the extremities were treated."

But Mondino was a dissector in name only: he did not perform the actual operations on the flesh. Three people, each with his own particular set of skills, did the real work. A barber made the initial cuts using sharp razors and scissors; a demonstrator pointed to each organ, extremity, or nerve as he removed it from the body; and the anatomist sat apart from the body and, never touching the corpse, explained to the masses what they were seeing. Like his colleagues, Mondino published a textbook, *Anatomia*, which became the go-to guide for dissectors for centuries afterward.

IN THE SIXTEENTH CENTURY BOLO-gna was awed by the dissections of a new anatomist, the Belgian Andreas Vesalius. His fame had come in part because while performing his own dissections he recognized that many of the earlier references by Claudius Galen had been faulty.

Vesalius had begun by dismembering dogs, kittens, and other small animals, not unlike Galen, but he later moved on to humans and was able to understand that the notions of all dissectors and anatomists—Galen's idea of the

Andreas Vesalius. Portrait of the famous anatomist renowned not only for his work but for his body snatching in the French cemeteries.

four humors—were wrong. This idea had developed in Galen's youth, when he declared that the body's organs possessed four humors, or substances: blood, phlegm, yellow bile, and black bile.

When these substances were not in accord with one another, according to Galen, disease and corruption occurred. For example, an excess of black bile would cause a person to show signs of depression and melancholy; of course if the person appeared sickly and pale, there must be an excess of the yellow bile. But Galen had not performed dissections on humans; instead he had performed vivisections of animals, which he believed were comparable to humans. For example, he thought human livers had five lobes, just like dogs', and that the human heart had two chambers, when it actually had four. Despite these incongruities, for centuries thereafter anatomists adhered to his ideas.

In early 1537, the custom of publicly viewing the dissection of dead bodies was reinstated after a hiatus of nearly eighteen years. Vesalius was given the honor of being the first anatomist to conduct the ceremony, and he needed to find a perfect specimen. With the lectures nearing, on a blustery winter evening, he heard that a thief had been burned at the stake. Body snatching was frowned upon, but Vesalius needed to get his hands on corpses so he could study his theories.

Disregarding the laws, Vesalius neared the burning spot and contemplated what was left of the smoldering corpse. He then climbed atop the stake and dismantled the remains, which came apart with a great creaking sound. "The brigand had provided the birds with such a tasty meal that the bones were completely bare and bound together solely by the ligaments," he later wrote. He went on, detailing how he had "pulled away a femur from the hip bone," and how, when he "pulled at the upper limbs, the arms and hands came away bringing with them the scapulae."

Making sure no one was watching this dreadful ordeal, he lugged the corpse away. "I allowed myself to be shut outside the city at nightfall; so keen and eager was I to obtain those bones that I did not flinch from going at midnight among all the corpses and pulling down what I wanted," he said, speaking of that experience and of others he had been involved in. "I had to climb the stake without assistance, and it took a great deal of effort and hard work. Having pulled down the bones, I took them away a certain distance and hid them in a secret place, and brought them home bit by bit the next day through another of the city gates."

Frontispiece to Vesalius's *De Humani Corporis Fabrica*. His most famous work, it was published in 1543, when he was in his late twenties.

This was not the only time he stole a corpse. While a student in Paris, he had often scavenged the cemeteries for bones and flesh, particularly at the Gibbet of Montfaucon, the notorious spot where criminals were hung. The cadavers were then hauled to his home, carefully stripped to the bare bones, boiled in large odorous vats of water, and cleaned. Then, using the cleaned pieces, he reconstructed the human skeleton. This practice of becoming such a hands-on dissector allowed him, at the age of twenty-six, to write the seminal work *De Humani*

Corporis Fabrica, "On the Fabric of the Human Body," a dramatically original work published in 1543. In short order, he became the finest and most sought-after dissector Europe had ever seen. In December 1537, he had been appointed professor of surgery at the famed University of Padua.

In 1540 officials from the University of Bologna invited Vesalius to give several lectures and anatomical presentations. Not surprisingly, thousands of students attended, though the most reliable note-taker appears to have been Baldasar Heseler.

"The anatomy of our subject was arranged in the place where they used to elect the Rector Medicorum," Heseler's notes read. "A table, on which the subject was laid, was conveniently and well installed with four steps of benches in a circle, so that nearly 200 persons could see the anatomy. However, nobody was allowed to enter before the anatomists, and altogether, those who had paid 20 soldi. More than 150 students were present, and D. Curtius, Eigius, and many other doctors followed Curtius. At last, D. Andreas Vesalius arrived, and many candles were lighted, so that all should see."

The wisps of smoke that arose from the melting candles undulated by the cadaver's feet and hands and rose above the dead flesh. Some spectators followed those tendrils skyward, but only for a few minutes, before they dispersed as they made it all the way to the painted ceiling. Then the students and spectators returned their eyes to the corpse and watched as a confident Vesalius worked on "the body cut up and prepared beforehand, already shaved, washed and cleaned." He pointed to each organ, bone, nerve, vein, referring to Galen's claims, refuting them, as the crowd stood by, either awed or insulted.

By the time Galvani took the stage in the 1700s, some of the practices normally associated with the lectures and demonstrations had been done away with—such as the clear division of labor—though the theatricality still thrived. The historian William Brockbank noted that the theaters had been constructed less for practicality than to merge art and science, a way for the public to take part in the demonstrations that for so long had held an air of secrecy and mystery. "It is clearly connected not only with the history of medicine and of teaching, but also with the history of art. The theater arose out of the stream of ideas which flowed through Italy at the time of the Renaissance. Its purpose was to offer a performance," Brockbank argued, "for an anatomical dissection those days was really more of a theatrical occasion than a lesson. The outstanding personalities and authorities of the town were invited to be present. It was the first laboratory, the first place where scientific research was carried out."

For much of Galvani's earlier career, he had been intrigued by other areas of medicine, including the study of the bone structure, the uterus, and particularly the development of the ear canals in humans and birds. But as time passed, he seemed less interested in those research subjects, especially after a fellow medic and researcher appeared to have used some of Galvani's publications on ear canals in his own text. A scandal broke out, but Galvani did nothing about it—either because he didn't relish such a public display of anger or perhaps because he'd become so absorbed by the study of animal electricity, he'd lost his passion for any other medical interest.

Whatever the reason, the seriousness of his behavior and his steady nerves came in handy in 1791. On March 27, he published his findings on animal electricity in the scientific journal of the Bologna Academy and the Institute of Sciences. *De viribus electricitatis in motu musculari commentarius* (Commentary on the effects of electricity on muscular motion) was the culmination of nearly eleven years of experiments. Many in the academic world read Galvani's work with interest, and though impressed, they were not greatly disturbed because Galvani was not the first, or the earliest, scientist to look into animal electricity. Two other Bolognese before him, Floriano Caldani and Giambattista Beccaria, had been able to elicit twitches from dead frogs. Galvani mentioned their findings in the *Commentaries*. But unlike Caldani and Beccaria, Galvani's scope was so massive, his research and experimentation so extensive, it was difficult to set it aside without giving it further thought.

Galvani spoke of the moment when all of his experiments culminated: "Accordingly, on an evening early in September 1786, we placed some frogs horizontally on a parapet, prepared in the usual manner by piercing and suspending their spinal cords with iron hooks. The hooks touched an iron plate; behold! A variety of not infrequent spontaneous movements in the frog. If, when they were quiescent, the hook was pressed with the finger against the iron surface, the frogs became excited almost as often as this type of pressure was applied."

He was showing that the frogs'

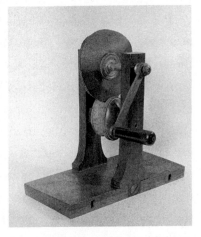

Apparatus formerly used by Luigi Galvani. This is a small plate electrostatic machine, used to harness electricity. One of the many objects found in his laboratory.

muscle contractions were the result of the vital fluid that circulated within their bodies. This fluid was then instigated to revitalize by a metallic arc that touched the crural nerves and muscles. The fluid then became excited, so to speak, which caused the movements in the frogs' limbs. Galvani's nephew Giovanni Aldini said this was something "no one had attempted" before. He came to believe that Galvani's "ingenuity" led to an understanding that "we may have muscle disposed to contraction by mere passage of a spark."

Throughout Europe, from France to England and Germany, scientists began replicating Galvani's experiments.

At the University of Pavia, Bassiano Carminati, a professor of anatomy, began to scrutinize a copy of the *Commentaries* sent to him by Galvani. Carminati was a well-known professor whose acquaintances included a wide range of colleagues and students, one of whom was the renowned physicist Alessandro Volta. Eventually Carminati passed the *Commentaries* on to Volta, who had kept abreast of the latest so-called discoveries in animal electricity, as well as the scientists who delved into the field.

Ordinarily, Volta never considered animal electricity a valuable source of scientific study. But as he reclined at his desk in Pavia reading the *Commentaries* as Carminati had urged him to do, he changed his mind. He was intrigued by Galvani's proposals and even considered dabbling in these studies himself. In March 1792, he connected the muscles of a frog, dismembered in Galvani's style, to its exposed nerves, without the use of the so-called artificial electricity. Surprisingly, he interpreted the results in the same manner Galvani had. He even announced, "We have to adapt to the idea that animal electricity exists." But he still had no real enthusiasm for the concept. He thought there was something odd in the experiments and doubted his findings. Even more, he doubted Galvani's.

Not long after Volta published his statements, Galvani received a letter from Bassiano Carminati that offered false praise and a certain measure of insult. Volta, Carminati explained, had been delving deeply into the subject of animal electricity, conducting experiments similar to Galvani's. But, as it stood, Carminati continued, Volta had also managed to disprove Galvani's claims, having "concluded that the deficiency of electric fluids exists on the part of the nerves . . . therefore, our distinguished Signor Volta, wishes the contrary of your opinion, which is not yet held as a settled thing."

This was a blow to Galvani. His findings, nearly eleven years of detailed experiments and studies, not to mention the thousands of frogs he had butchered, were being rebuffed in a matter of months. This was very much in keeping with Volta's character.

ALESSANDRO VOLTA'S UPBRINGING HAD NOT GIVEN ANY INKLING that he would go on to do great things. True, in 1800 he did construct the so-called voltaic pile, the first electric battery, which he created by further developing Luigi Galvani's ideas. He arranged two metals and a piece of fabric immersed in brine on a circuit, which then sparked an electric current. A few days later, he expanded the metals and alternated the zinc with copper, which produced a higher conductivity. He connected those with a wire and then noticed electricity flowing between the newly named voltaic pile and the wire itself. This new discovery revolutionized science in the nineteenth century. The medic Anthony Carlisle, William Godwin's friend, used the pile to experiment with water. And even Humphry Davy fiddled with it.

He had been born into aristocracy in the northern Italian city of Como, in the Lombardy region. His father, Filippo, died when Alessandro was eight years old—not that his death made that much of a difference. Even when alive, Filippo's propensity for stepping in and out of his children's lives made him seem like a semiabsent parent. However, his passing did cause financial instability for the family. Alessandro's mother, Maddalena, came from Italian nobility, but her title was in name only. When she became a widow with a brood of children, the family nearly sank into destitution.

Luckily, in 1756 the family learned that an uncle had died, bequeathing them a sizable inheritance. They suddenly found themselves owners of various properties, parcels of land in and around the vicinity of Como and a number of large and small homes, which they rented out, along with some estates where they could live. The money also allowed them to pursue a life of leisure, the indulgence of desires, even some that were wicked. As Alessandro grew from a boy into a teenager, those closest to him noticed the striking similarities between him and his dead father. Like his father, Alessandro had a taste for the good life; he liked to party. He could always be found enjoying himself at concerts, operas, feasts, any occasion where the liquor flowed freely and the women were willing and able.

Also like his father, he attended a school run by the Jesuits, though he quit after barely a year. Along with many young people of his age and social standing, he was attracted to the literary and scientific worlds.

Given his family connections, it was relatively easy for him to find patrons who would advance his aspirations. He also had an odd combination of ambition and conformity: he was willing to adhere to a particular philosophy if his own needs were met. In

the early 1760s—the author Giuliano Pancaldi says it was between 1762 and 1764—Volta wrote a prose poem of 492 verses
in which he tried to bring scientific rationale to those phenomena
for which rational meaning had yet to be found, such as lightning.
This not only sparked an interest in his own and others' literary works, it also brought about a desire to begin corresponding
with those whose interests matched his own, scientifically, but also
literarily—the so-called natural philosophers.

One such man was Giambattista Beccaria, whose favor and
input Volta wanted and needed. Despite Volta's numerous letters, the older man refused to indulge him. Still undaunted, Volta
continued to seek assistance from scholars who were not well-
known but who would give him opportunities for advancement.
In so doing his facility for social interaction developed—what one
might call the simple ability to flatter people in the right positions—a quality that garnered him numerous opportunities as a
lecturer and speaker, as well as interactions with various ladies of
important means, to whom he attached himself.

He was known for his charismatic personality, which he honed
carefully during years of social activities as well as his numerous
lectures. This gave him the upper hand during the controversy
with Luigi Galvani. Unlike Galvani, who came across as reserved,
or worse, even cold, Volta had a gregarious demeanor that put his
colleagues and acquaintances at ease. His knack for showmanship
helped his pursuit not only of peers, but also of patronage.

By the time of Galvani's *Commentaries*, Volta had been teaching
at the University of Pavia since 1778, with electricity, not animal
electricity, at the forefront of his career. But the reason Galvani's
experiments changed his mind, most especially, was because he

became convinced that Galvani had made a mistake, a somewhat obvious mistake.

Was it possible, Volta wondered, for a frog to act as a conductor? As his experiments continued, he came to realize it was not the frogs that possessed the so-called vital fluid, as Galvani proclaimed. What actually made them twitch was the metal of the outside apparatus that Galvani used, which then came into contact with a humid body. The metal produced the actual twitching, not the frog.

The scientific community and those who knew both parties geared up for a scientific showdown between two great minds, each eager to prove his own theories and disprove the other's. But it was not to happen.

On December 4, 1798, while the controversy was still in full swing, Luigi Galvani died, seemingly taking with him the notion of animal electricity, because Volta and his supporters had shed great doubt on Galvani's findings. By this time, Galvani had become a broken man. Lucia, his greatest supporter, had also died, and her death sapped the enthusiasm he had always possessed for his studies and research. In addition, the academic world in which he had been enveloped for decades was also in turmoil.

Along with a handful of other professors, Galvani had refused to swear allegiance to the constitution of the Cisalpine Republic, which meant he had been stripped of all his academic duties and honors. His nephew, Giovanni Aldini, had tried to persuade him to reconsider his stance, but Galvani refused to back down. Eventually, thanks to Aldini, Galvani was given the title of emeritus professor, though by then it was too late. Still, upon Galvani's death, Aldini decided he would restore Galvani's ruined reputation.

IN EARLY JANUARY 1802, A RESPECTABLE CROWD OF SCHOLARS AND doctors gathered in a semidarkened laboratory in Bologna, staring at the head of a slaughtered ox. They shifted uncomfortably in the unusually chilly room, rubbing their hands vigorously. Winter in Bologna could be cold, and this year was no exception. They shuffled in their chairs as they viewed the ox's head, the dead stump resting in the middle of a surgical table. Earlier that day, Aldini had procured a voltaic pile, which he hoped to use on the animal.

Aldini entered the room and stood by the table where the ox's head lay. He didn't feel the need to speak or to explain his doings, but with the flourishing gestures that were his stamp, he picked up an electrical arc, which sizzled loudly in his hands. Then he applied it to the ox's head. Right away the eyes flew open and the animal's ears began to twitch. He heard the crowd inhale sharply as the head, despite being decapitated, seemed alive, its "tongue . . . agi-

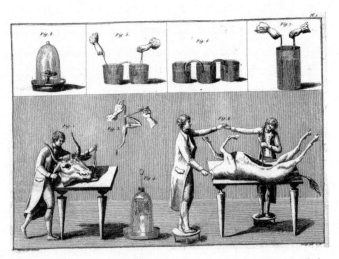

Giovanni Aldini performing galvanism experiments on warm-blooded animals, detailed and printed in his book *Essai théoretique et expérimental sur le galvanisme*, published in Paris, 1804.

tated . . . ." As Aldini later wrote, the ox moved "in the same manner as . . . the living animal when irritated and desirous of combating another of the same species."

Aldini knew his uncle had wanted his experiments to help find a cure for illnesses, particularly paralysis. While working at Sant'Orsola, Galvani had been keen to "subject an amputated arm or foot to his experiments," and "when a metal arc was carried from a muscle to the nerves, vigorous contractions suddenly arose." Often, Aldini had been employed as an assistant in his uncle's laboratory, where he had been exposed to a treasure-trove of amputated limbs, dead frogs, and other small animals in various states of decomposition, as well as assorted chemicals.

To continue his uncle's work, Aldini needed to properly master the voltaic battery. At first, he had a tendency to turn up the switches and knobs to such a powerful level that he inadvertently caused the animals' heads to convulse in a more repugnant manner, or to explode altogether. Sometimes, unaware of what was happening, the eyeballs rolled back and forth or protruded completely out of the animal's sockets—something he discovered when screams arose in his audience.

In January 1802, Aldini secured the bodies of two criminals who had been executed that morning. In life the two had been "very young . . . and of a robust constitution," traits Aldini always favored.

The visitors who arrived at the laboratory that night were expecting an ox, a dog, a calf, maybe. They were taken aback when Aldini uncovered two tables: on the first rested the lower body of one of the cadavers, and on the second was the second cadaver's separated head. Holding a zinc pile in his hands, Aldini neared the man's head, whose ears he had dampened with salt water. Then he

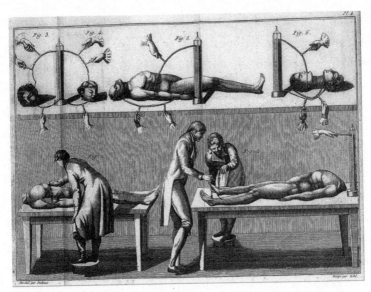

Galvani's experiment. From Giovanni Aldini's text *Essai théorique et expérimental sur le galvanisme*, depicting two decapitated cadavers and his efforts to restore movements to them.

"formed an arc with two metallic wires, which proceeding from the two ears, were applied." Those present moved uncomfortably in their chairs when, as he later wrote, he "observed strong contractions in all the muscles of the face, which were contorted in so irregular a manner that they exhibited the appearance of the most horrid grimaces. The action of the eyelids was exceedingly striking, though less sensible in the human head than in that of the ox." Despite the grimaces, some looked closer, as the corpse seemed to drool a thin rivulet of saliva from its lips.

To further prove the efficacy of the experiment, Aldini set the two criminals' heads side by side on the table. The official communication between the two heads was formed when the arc was stretched from "the right ear of one head, and to the left ear of the other." As he had anticipated, the two faces contorted as if alive, giving away such "horrid grimaces," those in attendance were "actually frightened."

Professor Mondini, a local academic and practitioner of medicine who had been eager to see for himself the result of such operations, pressed Aldini to repeat the experiments on different cadavers. January 1802 seemed to have been a prolific month for the executioner, as several criminals had their lives ended by the axe. This gave Aldini a steady supply for his experiments.

In front of a selected group of individuals, including well-known doctors and physicians, a very strong and muscular body was galvanized. "By applying the arc . . . ," Aldini later reported, "the violence of the contractions was much increased. The trunk was thrown into strong convulsions, the shoulders were elevated in a sensible manner, and the hands were so agitated that they beat against the table which supported the body."

Aldini knew that a number of scientists wondered why he was not using his experiments to discover cures for the sicknesses of the living. But he rejected this criticism and stuck to the notion that he was "promoting the welfare of the human race, and may be of service to cases of apparent death."

ON MAY 17, 1801, A YOUNG MAN NAMED LUIGI LANZARINI HAD BEEN institutionalized at the hospital of Sant'Orsola for what doctors had termed "melancholy." He was twenty-seven years old and a farm worker. Prior to his hospitalization, he had suffered from a prolonged bout of fever, which the doctors assumed had triggered his descent. Upon arriving at Sant'Orsola, Lanzarini began to complain of mistreatment and soon to display uneasiness around his doctors, so much so that his caretakers came to believe the melancholy had "degenerated into real stupidity."

Giovanni Aldini had always been intrigued by the notion of melancholy madness. He had come to believe the disease was due, first and foremost, to an imbalance in the brain, but also thought the responsibility could lie with accidents where the victims were hit over their heads, diseases of the body that traveled to the brain, or the bludgeoning of the victims. Such dire traumas to the skull and brain would often allow for "variations of the intellectual powers." He was determined to try the galvanization process on those suffering from melancholia madness, those who had "little hope . . . of their being restored to society." Lunatics, as they were called.

Prior to starting the experiments, he had tried to galvanize himself in a "long series of painful and disagreeable experiments." But even having had the experience himself did not stop him from trying galvanism on others. To Aldini, this was an extension of his uncle's desires.

Aldini conferred with Professor Gentilli and Professor Palazzi, both of whom were working at Sant'Orsola, and together they observed firsthand the tragic young Lanzarini. His melancholy had deepened, and the doctors were running out of options. Aldini suddenly realized he'd found the perfect subject: he was young; relatively healthy, aside from his current disease; and had only become a "lunatic" following a tangible ordeal, which was his fever. Aldini decided, with permission from Lanzarini's doctors, to administer galvanism.

During the first session, Aldini's therapies were administered slowly, adjusting the voltaic battery as he went along. He saw no marked improvement. Lanzarini was questioned about his disease and where he believed it had been generated. But his state of mind had been so damaged that it did not allow for any clear answers.

He merely stared at his doctors and Aldini, as well as the galvanization machine. His eyes were fixed and he slurred, which appeared to "indicate a great degree of stupidity and derangements."

As Aldini continued his ministrations, Lanzarini seemed to get better, his melancholy lessening. Aldini became convinced that his doings were able to "prove that Galvanism absolutely exercised an action on such a disease." Lanzarini didn't suffer from the process itself, and when doctors spoke to him days later, he was able to answer them—in a clear and concise voice—that he felt no adverse reaction to the voltaic arc Aldini was using. Over the following days, Lanzarini's pain evaporated, and to everyone's surprise, he even began to smile. He ate well and recovered enough that the doctors felt comfortable releasing him.

But Aldini was not entirely done with Lanzarini and invited him to Aldini's house. They ate and talked, and it was during one of these conversations that Lanzarini revealed that his father, Fabiano Lanzarini, had also been gripped by the same disease. He had been institutionalized at Sant'Orsola, where he had died on June 12, 1790. Aldini researched the information and learned it was true.

Knowing about Lanzarini's father didn't seem to change Aldini's beliefs. He did not make a connection between heredity and Lanzarini's current state of mind, but rather continued to believe that the environment had played a larger role in Lanzarini's disease, going so far as to suggest Lanzarini's moving away. "I advised Lanzarini to spend the rest of his life at a distance from his native country," Aldini wrote, "lest, having continually before his eyes these objects which had occasioned his disease, it might occur with double violence."

Lanzarini heeded his advice, but only for a short while. Suffering from "nostalgia," he later returned to his hometown, where, Aldini learned, he was doing well.

Lanzarini's case proved to be one of the few positive ones. The galvanization process seemed worthless for others suffering from mental disorders. Aldini even discovered that the process could be "dangerous" in patients where the disease was even more severe.

THOUGH ALDINI TRIED TO WORK WITH THE LIVING, HE CONTINUED most of his experiments on the dead. And performing galvanizations on dead corpses, he finally came to believe that "the heart, which, according to Haller's principle, is the first muscle that receives life and the last to lose it, in comparison of the other muscles, can with difficulty be made to feel the influence of the galvanic action."

Though he had no tangible proof it could be done, his new goal was to restart the heart and bring a dead corpse back to life. But the cadavers he got in Bologna were decapitated and long emptied of the vital force. He wanted to experiment on corpses that were still intact and were nearly warm to the touch. He needed to go someplace that offered such opportunities. England, with its progressive thinking, was the logical choice.

Chapter 3

MAKING MONSTERS

To Examine the cause of life,

we must have recourse to Death.

MARY SHELLEY,

FRANKENSTEIN; OR, THE MODERN PROMETHEUS

GIOVANNI ALDINI ARRIVED IN LONDON IN THE WINTER
of 1802 with his uncle's galvanic instruments and
a dash of the swagger that made him so controversial in
Bologna. His goal was to find a perfect dead body that he
could use to perform his galvanic experiments and prove
that his theories, and those of his uncle, were correct.

The city that greeted him was by then growing rapidly. By
the early 1800s, nearly one million people had poured into the
metropolis from all corners of Britain and abroad, allowing for
an amalgamation of cultures and social classes to take place.
In the West End the ostentatiously rich made use of every
technological advance available, every medical discovery money
could buy them, and every frivolous fad they believed would
prolong their lives, or at least rejuvenate their sagging bodies.

Meanwhile, the poor barely survived in the ramshackle
alleys of the old city. Those who had left the country and
come to the city to better their lives instead found filth, disease,
and destitution. The streets around them reeked with every
possible shade of humanity hawking their goods—shoes, pots
and pans, roasted chestnuts—all for a few measly coins. The
fetid, ripe smells that rose from the carts they dragged and the
stink wafting from the workhouses and spewing out from the
hundreds of chimneys mingled with the horse dung strewn across
the cobblestone streets, giving everything a foul, dingy quality.

The city was also at the cusp of a medical revolution. The main
hospitals were teeming with doctors and surgeons

experimenting on the living as well as the dead, busy trying to find new ways to cure the sick. And in discreet and out-of-the-way corners of the city, private laboratories had sprung up where doctors undertook experiments that hospital officials found too gruesome to perform in a respectable environment. During this time of medical innovation, there were charlatans, or so-called visionaries, who claimed their intentions and advances could save humanity. Unfortunately for Aldini, one of London's most scandalous quacks had experimented with electricity not long before he came to England.

James Graham, like most future doctors in England at the time, began his medical studies in Edinburgh, Scotland. But the curriculum did not include the new innovations in the medical sciences, so he traveled to the United States. He spent his early twenties in Philadelphia, where he learned about Benjamin Franklin's experiments with electricity. The notion of using electricity for medical purposes struck Graham like a lightning bolt and convinced him that electricity could be the cure for all that pained humankind. It has been suggested that in reality he never truly believed that electricity was a medical cure-all. Nevertheless, it was important to him that *others* would believe it.

When he returned to England, he was called upon by members of London's upper class to administer bolts of electricity; they believed it would cure them of headaches, menstrual cramps, gout, and everything else in between. With his reputation bolstered, he sought out a place where people could make use of his electrical machines, and in 1779, he opened the doors to the Temple of Health.

His clients were more than willing to pay the two guineas he charged for admission; once inside, their bodies and senses were immediately stimulated. To some, the opulent décor of the temple's rooms seemed entirely obscene: the shimmering, imported silks; the strong furniture whose curvatures implied, in not-so-subtle a way, the female body; the mirrors in which one could see oneself in all manners of seductive poses. The delicate perfumes and welcoming scents of lavender and roses that were misted in the air immediately recalled a fine spring morning and played delicately on the senses. Music flowed from unseen corners of the room, and tall Greek statues of voluptuous women stood seeming to guard the place. To the men's delight, scantily clad women frolicked among the statues.

By the time the clients saw the electrical apparatus, their bodies and senses were already stimulated and ready for anything. And for an extra fifty guineas at night, if they so desired, patrons could make use of the infamous "Celestial Bed," which had bolts of electricity crackling through it. Those who slept in the bed would be "blessed with progeny. Sterility and impotence would be cured." Though the environment was erotic, it was the electricity that did the actual work. It ran continuously from the headboard across the length of the bed, "filling the air with magnetic fluid calculated to give the necessary degree of strength and exaction to the nerves."

The Celestial Bed and the Temple of Health appealed to people who had latent issues with their sexuality—husbands who wished to step out on their wives in a comfortable environment, and wives who sought that extra something their husbands could not provide. Existing under the guise of a respected medical institution gave it credibility.

Though many attested to the benefits of Graham's electrical apparatus, others saw the temple for what it was: an upper-class

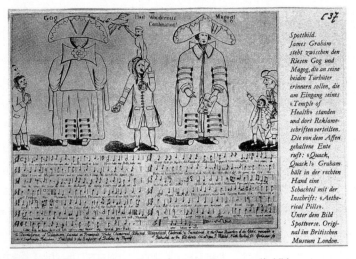

Caricature of James Graham (*center*). James Graham was called "the quintessence of quackism," the highlight of his career culminating in the building of the Temple of Health. There he made use of electricity to cure his sexually depleted patients. Here he stands in between Gog and Magog, the two footmen at the entrance of the Temple of Health.

bordello. Horace Walpole, who went to the temple for a session, described it as "the most impudent puppet show of imposition I ever saw, and the mountebank himself that dullest of his profession. A woman, invisible, warbled to clarinets on the stairs. The decorations are pretty and odd, and the apothecary, who comes up a trap-door (for no purpose, since he might as well come up the stairs), is a novelty. The electrical experiments are nothing at all singular, and a poor air-pump, that only hurts a bladder, pieces out the farce."

Faced with Graham's legacy, Aldini had to be careful how he promoted and described his methods. Graham had tried to restore peoples' depleted sexuality; Aldini wanted to restore life to the dead and to show that it was possible to do so with scientific methods. It was a highly dangerous move, impractical in a sense, and fraught with difficulties, not the least of which was the possibility of his experiments being undermined and of his being seen as a fool. He needed to do things precisely and in order.

First, he needed the perfect subject to galvanize. Youth and health were priorities, but most people died of disease and malnutrition, so it would be hard to find a vigorous body in its prime. He could rely on the gallows, as most anatomists did, but English law allotted only ten or twelve bodies for anatomizations, and those came highly prized by the medical institutions, which clamored after them. He could have hired a "resurrectionist," or a body snatcher, like most others in his position did, but they were expensive. Moreover, resurrection men were indiscriminate in whom they dug up—whether young or old, man or woman, healthy or diseased. The only requirement for a resurrectionist was a recent time of death, because then putrefaction would not have set in. Although Aldini needed a nonputrefied body, he would not stoop to employing a resurrection man to find one.

Most anatomists were not as picky as he was. Up until the 1800s, most medical students learned surgical procedures by watching their teachers perform actual surgeries. But as the population grew, students and doctors realized a more hands-on method was needed, one in which each student could work on a corpse. This hands-on practice, called the Parisian method, gave them a chance to study how each organ, muscle, nerve, and bone operated. It was called the Parisian method because the law in France allowed surgeons to use the unclaimed bodies left in hospitals and death houses. However, in England the gallows were the only *legal* source from which to collect corpses.

But the graveyards held a plentiful supply of unused corpses, and stealing them was not a serious crime. Actually trespassing was the grave robber's biggest concern. If anatomists and their cohorts (mostly students) were found lurking in a place where they did not belong, say a cemetery, they could be arrested. They could also get in trouble if they were caught stealing objects—for instance, anything that belonged to the corpse, such as the clothes he or she had been interred with, any jewelry, shoes for the girlfriend, or mementoes, like portraits. To be careful, they conducted all transactions during the most absolute stillness of the night, and if the moon only rose to a sliver, so much the better, because too much light could be devastating. But some learned that too much silence could also cause problems, as it was hard to disguise the echo of their footsteps running down the street with a corpse bouncing off their backs.

Snatching bodies proved a lucrative enterprise for the anatomists, though many found the actual digging of the bodies too stressful on their psyche. The general population thought it was disgraceful and gruesome. Doctors could have their reputations tarnished if they became known as grave robbers and body snatchers. Middlemen were needed to do the work for them.

The money the resurrection men earned was appealing. A corpse could yield more than a week's pay at any menial job. There was something in the forbidden act itself that was alluring to many as well. Christian Baroent described his work by saying, "The time chosen in dark winter nights . . . A hole was dug down to the coffin, only where the head lay—a canvas sheet being stretched around to receive the earth, and to prevent any of it spoiling the smooth uniformity of the grass. The digging was done with short, flat, dagger-shaped implements of woods, to avoid the clicking of iron striking stones. In reaching the coffin, two broad iron hooks under the lid, pulled forcibly up with a rope, broke off a sufficient portion of the lid to allow the body to be dragged out . . . the surface of the ground was carefully restored to its original conditions . . . the whole process could be completed in an hour, even though the grave might be six feet deep."

Resurrectionist gangs sprang up all over the city of London and the suburbs, working, for the most part, during the winter season. Their favorite haunts were the burial spots of the poor, who placed their dead in pine boxes that were easy to break into. If the gangs got particularly lucky, they found mass graves where several people who had died in the same day were buried together. The corpses were unearthed, put in sacks (hence the resurrection men's nickname "sack-'em-up men"), and dragged to the waiting party, most likely a servant working for a well-known anatomist in a back-alley laboratory, who would haggle over the price of the corpse as if it were a barrel of fruit.

Messr. Cruncher and his son, two typical men who worked as resurrectionists. These two were mentioned by Charles Dickens in his novel *A Tale of Two Cities*.

Ben Crouch was the leader of the most famous gang in that period. He was a foul-mouthed former pugilist whose physical strength was an asset when it came to digging out corpses but also to bullying others intending to enter the business. He also was a crook who would wait until his mates were drunk before dividing the take. With the advantage of sobriety, he managed to keep a larger share of the profit without anyone being able to tell. If someone pointed out the fact, the muscular Crouch didn't waste a minute but carefully landed a bejeweled fist (he was fond of wearing thick rings and bracelets) over the opponent's mouth, as if engaged in one of his former fights.

The other members included Bill Harnett; Jack Harnett; Tom Light; men named Daniell, Butler, and Hollis; and Joseph Naples, who might have been the only resurrection man who ever kept a journal that described all of their doings. It was not even a journal, but more of a log that told how many bodies they stole, where they removed them from, and where they sent them. Published under the title of *The Diary of a Resurrectionist*, it listed the gang's doings from 1811 to 1812, often with such simple entries as "Sunday, 21st, Went to S. Thomas's. Sent 1 to Mr. Tounton, 2 to Edinburgh S. Thomas's took 6 of the whole this week, came home and slept at home all night."

Naples was described as "a civil and well conducted man, slight in person, with a pleasing expression of countenance, and of respectable manners." He had learned to work briskly in the field and not to argue with Crouch, particularly when either one was intoxicated, which, according to the diary, was often.

There was no pretense that the corpses themselves had led lives, however difficult or distasteful, prior to their deaths. They were, according to Ruth Richardson, who wrote extensively about resurrectionist life, "bought and sold, they were touted, priced, haggled over, negotiated for, discussed in terms of supply and demand, delivered, imported, exported, transported . . . compressed into boxes, packed in sawdust, packed in hay, trussed up in sacks, roped up like hams, sewn in canvas, packed in cases, casks, barrels, crates and hampers, salted, pickled or injected with preservatives. They were carried in carts, in wagons, in burrows, and steamboats; man-handled, damaged in transit, and hidden under loads of vegetables. They were stored in cellars and quays. Human bodies were dis-membered and sold in pieces, or measured and sold by the inch." They were classified according to size, small being "a body under three feet long; those were sold at so much per inch and were fur-ther classified as 'large small,' 'small' and 'foetus.'"

Sometimes the public became aware of a particularly gruesome and horrific case. In 1826, someone was shipping three containers labeled "Bitter Salts" from the port of Liverpool to Leith, on the vessel *Latana*. What happened next was printed on a broadside, which read in part, "The casks remained on the quay all night, and this morning, previous to their being put on board, a horrid stench was experienced by the mates of the *Latana* and other persons . . . this caused some suspicion that the crates did not agree with their super-inscription which was 'Bitter Salts,' a constable was sent to the quay, and he caused the casks to be opened, where eleven dead bodies were found within, salted and pickled."

Bodies had become just objects and things. The living had very carefully removed all feelings associated with the dead. Aban-doning all scruples, as soon as the dead were dealt with and the business with the living concluded, the men suddenly found them-selves with money in their pockets. If they were working within a group, the money itself would not have been very much, and *if* diplomacy prevailed (which it almost never did), the earnings were split equally among all the members. Either way, the earnings were enough to pay for a pint at the local joint, likely the most famous gathering spot of all among resurrection men, the Fortune of War pub, epicenter of the resurrectionist's life.

This place just happened to be located near St. Bartholomew's Hospital and Medical College, and not too far from the Old Bailey courthouse. If one was keen to do so, a line could be drawn between the courthouse, where the criminals were hung; the pub, where the body snatchers hung out; and the hospital, where eventually the corpses were dissected. The men who frequented the pub kept abreast of the latest convicts in the Old Bailey; they knew whose time was coming up. And they were also aware of the doctors working in the hospital across the street and of their particular needs, requirements, and preferences.

The doctors had an uneasy relationship with the resurrection men. They needed their help to procure bodies, but they were appalled by their inhumane actions. Once they agreed to buy a body from a certain gang or resurrection man, they could not back down or change their mind. If the doctor was seen sneaking around searching for a better deal, the resurrection man would seek retaliation, and Ben Crouch and his gang were known to be quite vicious about their tactics. Sometimes they broke into dissecting rooms and destroyed bodies ready to be examined, or, more often, they called the police, ruining the doctors' reputations.

Joshua Brookes was one doctor who refused to follow the rules set up by the resurrection men. To his own detriment, he bought corpses from whoever offered the best deal. Gangs who expected loyalty often called the authorities to Brookes's laboratory, where trouble arose. Once rotting bodies were left outside his house, where two young women walking by early in the day found them and screamed, alerting the neighbors to what had happened.

Strangely enough, few of the doctors bequeathed their bodies to be used for dissections after they died. Several went so far as to purchase coffins that were being sold at enormous prices and were said to prevent the picks and shovels of the resurrectionists from breaking through. One such doctor, on the brink of death, imagined that his assistants would descend on his dead carcass like vultures and wrote a poem begging,

> *And my 'prentices will surely come*
> *And carve me bone from bone,*
> *And I, who have rifled the dead man's grave,*
> *Shall never rest in my own.*
> *Bury me in lead when I am dead,*
> *My brethren, I entrust.*
> *And see the coffin weigh'd I beg*
> *Lest the plumber should be a chest.*

Giovanni Aldini could have gone to the Fortune of War pub, or one like it, and engaged one of the resurrection men. He could have made his specifications known and perhaps one of them, in time, would have come up with the right subject, for a particular sum of money. But Aldini had another plan in mind. He not only wanted to find the perfect man to restore life to, but he also hoped to attract the right people who would back up his concepts, and possibly even pay for his stay in London. That's when he approached the members of the Royal Humane Society.

William Hawes and Thomas Cogan founded the Royal Humane Society in 1774. They were physicians who were concerned when they realized that a great number of people in the city's hospitals were being taken for dead when they were still alive. To make matters worse, some of those still-living patients were being buried alive. This frightened doctors, as they were the ones to declare the actual time of death, and made patients fear going to sleep one moment and waking up the next in a pine box. The society was initially called the Society for the Recovery of Persons Apparently Drowned, as restoring life to the drowned was their first order of business.

Unlike other doctors, they found it necessary to push the new technique—an unproven one, no less—of resuscitation. They came up with a list of objectives the society would aim for. They agreed that people at large would help them, for why wouldn't people support the art of resuscitation if it benefited them? Among their goals was the one they felt would attract the public the most: they would actually pay those who not only tried to bring someone back to life but actually managed to do so.

Aldini approached the Royal Humane Society with a solution. Unabashedly suave in his demonstrations, he had come to realize, unlike his uncle, that his spectators came to view the experiments as much for the grand shows he provided as for the potential outcome. He also hoped the men of the society would provide him much-needed validation for his contraptions, as well as support, and introductions to the even more refined society of London. While the society members found his manners and his propositions a little unusual and his self-assurance almost bordering on conceit, they nonetheless agreed that his methods were worthy of a try. They also agreed to help with the more tangible issue at hand: finding the right corpse.

WHEN GEORGE FOSTER WAS ARRESTED, HE WAS "INDICTED FOR THE willful murder of Jane Foster, his wife, and Louisa Foster, his infant child." This had occurred on December 5, 1802, in a canal at Westbourne Green, in the city of Westminster. Despite the dire accusations, Foster believed a grave mistake had been made and he would soon be vindicated. Undaunted and ignorant in the ways the laws worked, he believed that the testimonies of his neighbors, coworkers, the people he had lodged with in the past, and the ones he was now living with would set him free.

Jane Hobart, Jane Foster's mother, was the first to take the stand. A bundle of grief, anger, and resentment underneath her raggedy garments, she stared at Foster as she gave detailed accounts of how she had removed her daughter and granddaughter from the poorhouse to care for them. She unleashed her tongue as she began to tell the courts and those in attendance of Foster's desire to place even the smallest of his children, the now-dead infant Louisa, in a workhouse.

The testimonies that followed weren't any more in his favor. The people who testified said they remembered seeing George Foster and his wife and child arguing on the fourth. They had not been on loving terms and melancholy had plagued Jane. Worst of all, it appeared that George Foster had wanted to get out of the marriage.

When Hannah Patience, the keeper of the Mitre Tavern, was sworn in, she spoke of George, Jane, and little Louisa as having been at the tavern for a good while, where she had served them some liquor. Though they had sat drinking for some time, the matter didn't point to the fact that the two might have been drunk and that the liquor might have instigated a domestic dispute while on the road back to their place of sleeping, which in turn might have caused George Foster to commit murder against his wife and daughter.

The official case hinged on the testimony of Sarah Daniels, a girl who had seen the Fosters when she went to buy candles for her employer that afternoon and was the last one to see Jane and Louisa alive. But when Daniels took the stand and was questioned about what she knew and what she had seen, her testimony seemed rehearsed. Though she *had* seen the Fosters walking near the tavern, she had not witnessed the murder.

Next was John Atkins, who had found the corpses. On Monday, December 6, a bitterly cold day, he was walking near the canal when he discovered an infant girl's body under the bow of his boat. He was directed by the authorities to search the nearby waters of the canal even further, and on the third day, he and some other men found the body of a woman entangled in twigs and grass. They dragged it up. Atkins's description of the infant child's being dredged out of a frigid canal in the middle of winter, followed a few days later by her young mother, must have devastated the hearing's attendees, including George Foster.

Foster was arrested soon after the bodies were discovered. He had also signed a statement to which some more of his words were later added. Those read in part: "I left her directly when I came out of the Mitre Tavern, which was about three o'clock ... in order to go to Barnet, to see two of my children, who are in the work-house there; I went by the bye lanes, and was about an hour and a half walking from the Mitre to Whetstal; when I got there I found it so dark that I would not go on to Barnet, but came home that night; I have not seen my wife or child since; I have not enquired about them, but I meant to have done so tomorrow evening, at Mrs. Hobert's."

The coroner reported no bruises, blows, cuts, marks, or other injuries of any sort on the bodies. When he was asked what he believed about the crime, if indeed a crime had been committed at all, he shook his head and told the courtroom the deaths had been "accidental." He had come to the conclusion that the woman had fallen into the river, for "between the rail and the side of the river it is impossible to walk with safety, it is so slippery like soap."

Perhaps Jane had committed suicide. Mr. Alley, Foster's court-appointed barrister, asked witnesses if she had ever said anything about her desire to die.

Sarah Going shook her head. She had known Jane for some years before, when she had stayed with the Fosters. Sarah Going said no again. She had been so surprised by the Fosters' turn of events, their financial situation and marital troubles, for George Foster had seemed like a very good husband and father.

She must have been the only one to feel that way, for the Second Middlesex Jury before the Lord Chief Baron quickly returned a guilty verdict. George Foster would hang. Worst still, as he had imagined and feared, his body would be handed over to the anato-mists.

Giovanni Aldini had finally found his perfect corpse.

IN 1836, CHARLES DICKENS VISITED THE PRISON OF NEWGATE. BY then it must have occurred to him that an article detailing the prison's interior—the building, the prisoners, their doings, and how they spent their final moments—would entertain his readers. The dismal setting no doubt compelled him to visit, but he was probably also baffled by those who continually chose to attend the executions, because he made it clear from the beginning that he was writing for them. He wondered how it was possible that those moving to and fro about the city and in the vicinity of the prison could go by "without bestowing a hasty glance at its small, gated windows, and a transient thought upon the condition of the unhappy beings immured in its dismal cells."

He hoped to see those dismal cells as well as a glimpse of those unhappy beings inside who were on their way to the gallows. On more than one occasion he witnessed criminals being hung and used their lives and experiences in his books, such as *Bleak House.*

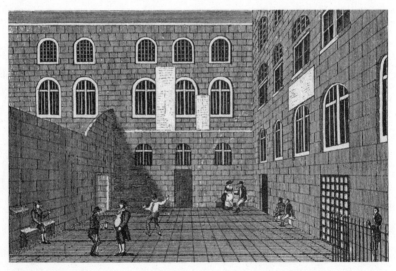

Newgate Prison's inner courtyard during the eighteenth century. It is here that most criminals where brought prior to their public executions, including George Foster, who underwent galvanism experiments at the hands of Giovanni Aldini. Charles Dickens, as did many others, passed by the prison on a daily basis, and he was prompted to write his famous essay "A Visit to Newgate."

But all in all, he found it deplorable that such occasions were often seen as amusing sociable gatherings, as if the passing of a human being at the hands of another carried with it a note of joviality.

"I did not see one token in all that immense crowd of any emotion suitable for the occasion," he wrote in a letter following the hanging of a famous prisoner. "No sorrow, no salutary terror, no abhorrence, no seriousness, nothing but ribaldry, debauchery, levity, drunkenness and flaunting vice in fifty other shapes. I should have deemed it impossible that I could have ever felt any large assemblage of my fellow creatures to be so odious."

On a gray and desolate day akin to those described in many of his novels, Charles Dickens walked around the prison. He was struck most by the cells of the condemned, where the prisoners stayed just prior to their execution. "It was a stone dungeon," he later wrote, "eight feet long by six wide, with a bench at the upper end, under which were a common rug, a Bible and a prayer book. An iron candlestick was fixed on the wall at the side; and a small high window in the back admitted as much air and light as could struggle in between a double row of heavy, crossed iron bars. It contained no other furniture of any description."

The institution was dilapidated and suffused by the odor of death and suffering. As Dickens looked at what was then an empty cell, he began to conjure the shape of a man, a doomed young man crouching upon himself. "Conceive the situation of a man," he wrote. "Spending his last night on earth in this cell . . . [having] neglected in his feverish restlessness the timely warnings of his spiritual consoler; and now that the illusion is at last dispelled, now that eternity is before him and guilt behind, now that his fears of death amount almost to madness, and an overwhelming sense of his helpless, hopeless state rushes upon him, he is lost and stupefied, and he has neither thoughts to turn to, nor power to call upon."

Much like the phantom criminal convicted and awaiting execution in Dickens's article, George Foster also found himself alone in his madness and fears of death. But his fears—as with those of the others condemned to die—must have gone beyond the actual moments on the gallows. Like many, he also must have feared eternity, as well as the possibility of not getting there at all.

In the Murder Act, passed in 1752, dissection had been added to the sentence of hanging for the specific purpose of inflicting "further terror and a peculiar mark of infamy" upon the criminal. The act, named "an Act for better preventing the horrid crime of Murder," had been drawn up because killings were on the rise in the city and Parliament wished to frighten people into keeping law and safety. It stated that "whereas the horrid crime of Murder has of late been more frequently perpetrated than formerly and particularly in and near the Metropolis of this Kingdom, contrary to known Humanity and Natural Genius of the British Nation: and whereas it is thereby become necessary, that for some further terror and peculiar Mark of Infamy be added to the Punishment of Death."

In addition, severe punishment was also extended to those who tried to help the criminal in any way. Because the hangings were in reality just slow strangulations (the necks did not break cleanly, but the convicted merely asphyxiated slowly), family members and friends were often seen running from the crowd toward the gallows and pulling on the men's legs to hasten their deaths and diminish their tortures. The Murder Act stated that if the family tried to rescue the body from the anatomists before or after death, they could be arrested and punished with transportation "to some of His Majesty's Colonies of Plantation in America, for the terms of Seven Years." If somehow they made it back to England before the seven years ran out, they could also suffer death by hanging.

A series of articles printed by the Royal College of Surgeons under the title *Echoes from the College of Surgeons* outlined the whole process in clear and matter-of-fact prose. Its readers found themselves repelled by the proceedings:

> *The executions generally took place at eight o'clock on Mondays, and the cut-down, as it is called, at nine, although there was no cutting at all, as the rope, with a large knot on the end, was simply passed through a thick and strong ring, with a screw, which firmly held the rope in its place, and when all was over, Colcroft, alias "Jack Ketch" would make his appearance on the scaffold, and by simply turning the screw, the body would fall down . . . On extraordinary occasions visitors were admitted by special favour. The bodies would be stripped, and the clothes removed by Colcroft as his valuable prerequisites, which, with the fatal rope, were afterwards exhibited to the morbidly curious, at so much per head, at some favored public house. It was the duty of the City Marshal to be present to see the body anatomized, as the Act of Parliament had it. A crucial incision in the chest was enough to satisfy the important city functionary above referred to, and he would soon beat a hasty retreat, on his gaily-decked charger, to report the due execution of his duty. Those experiments concluded, the body would be stitched up, and Reatison, an old museum attendant, would remove it in a large cart to the hospital, to which it was intended to present it for dissection.*

George Foster must have worried about the possibility of awakening during the act of dissection. He had probably heard stories of those who had not been truly dead when cut down from the gallows and had reawakened on the anatomists' table, just as the surgeon's knife was being plunged into their abdomen. If this had happened to him, he would have been taken to the gallows again, for a renewed hanging. In addition, there were religious connotations attached to the act of dissection. Most Londoners believed that, on the Day of Judgment, physical bodies would rise from the dead in order to meet the Lord. But if a body had been hacked to pieces and its remains scattered who-knows-where, that would be impossible. Thus, most criminals hated dissection for more reasons than could be realized: it would be the end of them not only in this lifetime but forevermore.

On the day of his execution, Foster confessed to the murder of his wife and daughter. His confession was later printed in the *Newgate Calendar*, the prison's bulletin, and everyone seemed satisfied with it, particularly Dr. Ford, the prison's priest, who now believed Foster could go to the other side with a clean conscience. It did not matter that the confessions were usually coerced and made only as an act to appease prison officials.

In the crowd, Mr. Pass, a somewhat unknown and shadowy character in the entire ordeal, waited for Foster's final demise. Mr. Pass was a failed surgeon who now worked as a middleman of sorts for the Royal College of Surgeons; nicknamed "the Beadle," he procured corpses for the anatomists whose experiments required certain services and secrecy. Not only had he helped Aldini in his pursuit, but he stood by as Foster's "cap was pulled over his eyes, when the stage falling from him, he was launched into eternity."

Prior to his death, Foster was asked if jealousy had led him to commit the "horrid act." He never responded. It was later reported that neither jealousy nor drink had been at play: Jane had jumped into the river, taking Louisa with her. But by the time those details were uncovered the deeds on Foster's body had been committed, and all of London and beyond had come to know of them.

"FIRST, THE FLUID TOOK OVER A LARGE PART OF MY BRAIN, WHICH left a strong shock, a sort of jolt against the inner surface of my skull. The effect increased further as I moved the electric arcs from one ear to the other. I felt a strong head stroke . . ." Giovanni Aldini was no stranger to galvanizing himself. He had tried such

HALL OF THE ROYAL COLLEGE OF SURGEONS.

The Royal College of Surgeons. Many anatomical experiments were conducted within the halls of this institution. Giovanni Aldini appealed to the members of the College in order to test his ideas on galvanism, but also for assistance in finding the perfect subject to galvanize.

experiments in Bologna, prior to doing so on the mentally ill. Now he repeated them again. Restarting the heart required precision, and he wanted to make certain everything was in order. He used himself before the January 17 date, which was to be the climax of all he had worked for.

As the law required, after Foster was executed, his body was left hanging for an hour. It was January, and the winter temperatures played havoc on Aldini's plans. He later complained about the condition of the body, which had been left dangling "in temperature two degrees below the freezing point." And after the noose was cut, the body was brought to a house, a small shed really, where it remained for a while longer, further cooling it down, which the courts required. Mr. Pass was not allowed to take possession of the corpse until the Royal College of Surgeons' necessary regulations had been followed.

Aldini surrounded the body, along with Mr. Keate and Mr. Corpue, officials from the college. They clustered around Foster as Aldini attached the first electrodes to the limbs, spots marked on the face, near the ears, and by the eyes where a solution of salt had been applied. Aldini powered his battery, which began to sizzle loudly, then "one arc being applied to the mouth and another to the ear . . . galvanism was communicated by means of three throngs combined together, each of which contained forty plates of zinc, and as many of copper."

The officials looked on as connections were made, and in a corner of the room, Mr. Pass watched, uttering not a word, not even when "on the first application of the arcs the jaw began to quiver, the adjoining muscles were horribly contorted, and the left eye actually opened." For those who had not witnessed such things before, Foster actually appeared to have returned to life and was now staring up at them.

Aldini had barely begun. He moved the arc to the ears, which caused the head to turn from side to side, again giving the impression that the corpse was taking in the crowd. Aldini gave more power to the battery, and the movements became more pronounced, so that "a convulsive action of all the muscles of the face" occurred, and "the lips and eyelids were also evidently affected."

But Foster was still dead. Despite Aldini's best intentions, the heart did not pump. As that had been the goal, he "endeavored to execute the action in the ventricles, but without success." He administered galvanism externally where the heart was located, but still nothing happened. As such, he concluded that if the arc were administered directly onto the organ, a reaction would be had. He cracked open Foster's ribs, "exposing the heart in situm." The heart, where the vital force, he believed, resided, now lay exposed before all to see. It rested, ready to take life anew.

Aldini reached for the arc and neared the surface of the organ, "then . . . the substance of the fibres, to the carnal columnae, to the septum ventriculorum, and lastly, in the course of the nerves by the coronary arteries." Stubbornly, Foster refused to reawaken. Aldini continued on, but all was "without the slightest visible action being induced."

For a second prior to opening the chest cavity, they had seen the chest heave up and down as if the corpse, to match their eagerness, had caught its breath. But in reality, nothing had occurred. The futile attempts continued for a bit longer, until they were given up altogether. Everyone was aware of the sobering reality that George

Foster had been dead at the gallows and he remained so after being galvanized. Disillusioned, they disbanded into the night.

ALTHOUGH ALDINI HAD FAILED, HE STILL BELIEVED IN HIS THEORIES. The battery had caused the botched galvanization, not him. "The troughs were frequently renewed, yet towards the close they were very much exhausted," he was later to write in his notes. "No doubt, with a stronger apparatus we might have observed muscular actions much longer."

One person who would not attend any more of Aldini's galvanizations was Mr. Pass. Shocked by what he'd seen, he died that evening, possibly of fright, as the *Newgate Calendar* explained. In reporting on Aldini's experiments and Pass's death, it was written, "On the first application of the process to the face, the jaws of the deceased criminal began to quiver, and the adjoining muscles were horribly contorted, and one eye actually opened. In the subsequent part of the process, the right hand was raised and clenched, and the legs and thighs were set in motion. Mr. Pass, the Beadle of the Surgeon's Company, who was officially present during this experiment, was so alarmed that he died of fright soon after he returned home."

Unbeknownst to Giovanni Aldini, his experiments, though flawed, would live on, albeit fictionally, in the hands of another scientist, Victor Frankenstein. It was Frankenstein who would successfully bring to life his own fiend.

Chapter 4

A MEETING OF TWO MINDS

Fillet of a fenny snake,
In the cauldron boil and bake;
Eye of newt and toe of frog,
Wool of bat and tongue of dog,
Adder's fork and blind-worm's sting,
Lizard's leg and owlet's wing,
For a charm of powerful trouble,
Like a hell-broth boil and bubble.

Double, double toil and trouble;
Fire burn and cauldron bubble.

WILLIAM SHAKESPEARE,
MACBETH, ACT 4, SCENE 1

IN 1502, A YOUNG WOMAN NAMED ELSA WALKED ACROSS the Devil's Bridge and hurtled herself into the frigid waters below. The bridge had a nasty reputation: located in a canton near Lake Zurich, it stood above a river that was so treacherous, many deaths often occurred there during the spring months of April and May, when the seasonal rains swelled its banks and caused it to overflow.

The bridge got its name from the tale of a local shepherd who one evening while tending to his flock found himself at the edge of the river, and, in trying to cross it, became so frightened by the violent currents that in a moment of passion he called not on God but on the devil to appear and build a bridge that would span from one end of the river to the other. The devil, of course, obliged the shepherd, but in true devilish fashion sought something in exchange for his favor: the soul of the first entity to cross the bridge, he said, would accompany him into hell.

The shepherd pondered this for a moment: true, he was grateful to the devil for the newly constructed bridge, but the idea of paying for it with his eternal soul did not really appeal to him. Nonetheless, he agreed to the devil's request.

The devil stood by as the shepherd began his walk across the bridge, but cunningly, the shepherd let one of his goats walk ahead of him; after all, the devil had not specified whether the entity had to be human or animal. The devil watched dumbfounded. Did a shepherd, a mere mortal, think he was going to get away with that? He looked about and found a large boulder that he could hurl at the bridge and smash it to pieces. But in that instant, a local woman passed by and saw what was occurring. She whispered a prayer toward the boulder, and it became so heavy, not even the devil was able to pick it up. The devil, having been conned by two humans, skulked back into hell.

Following young Elsa's suicide, her husband, Wilhelm, didn't want their nine-year-old son, Philippus Aureolus Theophrastus Bombastus von Hohenheim, to have any connection to the devil, so he moved the child away from the area. But not even two decades later, Philippus, who by then had learned the art of alchemy, thanks in part to his having been an apprentice at his father's side, sought knowledge beyond his father's purview and called on the devil himself.

Phyſick Proſſeſsor at Basil.
Philip Theophraſtus PARACELSUS *He died at Saltzburge An.° Dom: 1540. aged 47 yeares .*
W. Marſhall ſculpſit .

Paracelsus, famous alchemist said to have sold his soul to the devil in exchange for forbidden knowledge. He is also credited with having created the homunculus, the animated being made from bits and pieces of organic matter. Percy Shelley avidly read his works, as did the fictional Victor Frankenstein.

With that knowledge, Philippus became Paracelsus and began revolutionizing the discipline of alchemy through what he believed was a magical ability to transform metals into gold and, furthermore, to extend life or even to create it entirely anew. For some, Paracelsus represented everything that was successful about the art; he had developed the elixir of life and was capable of concocting a life-form from almost anything. But for others, as Charles Ponce, one of Paracelsus's biographers, noted, he was nothing more than "the Faustian archetypal quack, that notorious seeker of wisdom who sold his soul to the devil in exchange for forbidden knowledge."

In the early 1800s, the alchemical works of people such as Paracelsus were on the minds and reading lists of natural philosophers, certain scientists, and particularly the literary community at large. Like many Romantic poets, who hoped to find a higher self in nature, alchemists tried to change nature in order to understand the self and create life anew, both literally and figuratively.

This desire for higher knowledge did not occur by coincidence. The French Revolution had come and gone and had left in its wake two new movements: the dawn of the Industrial Revolution as well as the beginning of a scientific revolution. At this time, the very nature of what it meant to be a man, or rather a human, was being questioned. The notion of reawakening the dead with a bolt of electricity, and the experiments designed to do so, brought to light certain moral questions people had no definite answers to: Was man a creature created by a God who dished out values and properties according to his fancies? Or was man a machine powered by an internal galvanic fluid, which in turn could be sparked alive by a rush of electricity? Did man possess a soul endowed by God? Or was he merely a soulless automaton?

For members of the Church, this last idea was abhorrent. If people came to believe that man was soulless, God would be taken out of the equation. But for some members of the scientific community, not to mention owners of industries, the idea of soulless humans had merit: after all, if the dead could reawaken, or even better, be created entirely anew, then potentially a new race could be unleashed and commanded as one pleased.

These newly established thoughts were making the rounds

among the wildly popular salons, scientific circles, and literary communities. Pamphlets, books, and newspaper articles included discussions on the topic, with people responding with their own interpretations of what would happen if those ideas came to fruition. One of those responding was Mary Shelley.

In her 1818 novel, *Frankenstein; or, The Modern Prometheus,* Shelley has Victor Frankenstein create a soulless automaton from bits and pieces stolen from gloomy graveyards and death houses—and then bring that being to life by a force of electricity harnessed from the sky. But the monster of Mary Shelley's work eventually overpowers his creator and runs wild, destroying everything and everyone Victor loves.

People questioned where Mary Godwin Shelley, a teenager, found the inspiration for her fictional characters. Of course, the experiments of Giovanni Aldini, Luigi Galvani, and Humphry Davy were the topic of many conversations during that time. But what about the alchemical works of people like Paracelsus, Cornelius Agrippa, and Simon Magus? How did Mary Shelley know about them? Did she understand their implications? What their authors had done that brought their names into infamy?

Cornelius Agrippa, born Heinrich Cornelius Agrippa von Nettesheim, was accused of consorting with devils. His most famous work, *De occulta philosophia, libri tres,* was a practical guide to not only the occult but demonology. Legend suggests that on his deathbed, he allowed the black dog he always kept by his side to leave his house and roam the countryside free, a symbol of a demon being unleashed on the world. Villagers who came across the large animal thought this was a bad omen. Though Agrippa would eventually appear in *Frankenstein,* he first became famous in Goethe's *Faust.*

Simon Magus, on the other hand, was also known as Simon the Sorcerer. His exploits are little known beyond the apocryphal text the Acts of Peter, where Simon's battle against the apostle Peter are showcased. In a particular infamous duel that brought the sorcerer to the public's eyes, the two men faced one another in the Roman Forum, Simon performing what appeared to be a magic trick that allowed him to levitate above the crowd. Watching the event, Peter begged God for help in trying to stop Simon's flight. God listened and caused Simon to fall midflight. The crowd, who had enjoyed the performance until then, suddenly believed that Simon had been performing a parlor trick and turned on him, stoning him to death. It is this event that began the bickering between magic and religion.

Mary Shelley's novel begins with Victor Frankenstein describing the work of these alchemists: "When I was thirteen we all went on a party of pleasure to the baths near Thonon; the inclemency of the weather obliged us to remain a day confined in the inn. In this house I chanced to find a volume of the works of Cornelius Agrippa." Victor's father did not endorse his readings, calling them "sad trash," but Victor persisted. "I continued to read with the greatest avidity. When I returned home my first care was to procure the whole of this author, and afterwards of Paracelsus and Albert Magnus. I read and studied the wild fancies of these writers with delight; they appeared to me treasures known to few besides myself."

What did Victor Frankenstein find in those works that sent his mind aflutter? What captured his fancy and propelled him forward with his plan? Maybe it was the fact that each, in his own way, claimed to have found the elixir of life, with Paracelsus being the first to use the term *homunculus* in reference to the animated being he gave life to and that was made entirely of bits and pieces that he had stitched together. This little being, the homunculus, was actually the prototype for what Frankenstein was about to create on his own device.

THE RECIPE SEEMED SIMPLE AND STRAIGHTFORWARD: A SCATTERED pile of bones, some sperm, and "skin fragments and hair from any animal of which the homunculus would be a hybrid." All of these were the essential ingredients needed to create life. In his book *De rerum natura*—*Concerning the Nature of Things*—Paracelsus instructed the would-be alchemist to amass all needed supplies in a heap, knead the ingredients into a shape, and later bury the whole concoction in the ground, where, surrounded by horse manure, it would come to life after forty days.

"Let the semen of man putrefy itself in a sealed cucurbit with the highest putrefaction of the *vester equinus* [horse's womb] for forty days, or until it begins at last to live, move, and be agitated, which can easily be seen. After this time it will be in some degree like a human being, but," Paracelsus wrote, "nevertheless transparent and without a body. If now after this, it be every day nourished and fed cautiously and prudently with the Arcanum of human blood, and kept for forty weeks in the perpetual and eternal heat of a *vester equinus*, it becomes, henceforth, a true and living infant, having all the members of a child that is born from a woman, but much smaller. This we call a homunculus; and it should be afterwards educated with the greatest care . . . until it grows up and begins to display intelligence."

If creating life from scratch seemed too complicated, the alchemist could simply bring back the dead. This notion of resuscitation captivated the natural philosophers and Romantic poets of the 1800s. Resuscitation was possible, Paracelsus believed: "Death is twofold, that is to say, violent or spontaneous. From the one, a thing can be resuscitated, but not from the other."

It was always thought that Paracelsus learned the secrets of alchemy from his father, who, following their departure from Switzerland, settled as a physician in Villach, a German outpost. As his father's apprentice, Paracelsus was privy to all the secrets of the trade, which he later developed further and which eventually allowed him, critics believed, to create the being called a homunculus. The small creature, a mere foot tall at most, awoke suddenly and began "to live, move, and be agitated," until it stared glassy-eyed at its surroundings and its creator. But, much like in *Frankenstein*, problems soon arose when the creature, eager to further its knowledge, defied Paracelsus instead of bowing down to him. When word of this spread, it widened Paracelsus's reputation for performing the devil's handiwork. He was also an itinerant doctor, and unusual stories arose in the cities he visited across Europe. One prominent tale began to circulate in the German town of Ingolstadt.

Coincidentally, the character Victor Frankenstein attends the University of Ingolstadt following the tragic death of his mother. That's also where he meets the charismatic Professor Waldman; where his ideas about creating life come into play; where he raids the cemeteries; and where, finally, he brings the creature to life.

While at the university, Frankenstein finds himself in a peculiar state of mind. Bereaved, he ponders the unnaturalness, the evil the entire process of death and decay truly is. Having already read the works of the alchemists, Professor Waldman's teachings solidify his thoughts and the direction of his work.

Waldman does not scoff at the alchemists, and this allows Frankenstein to combine his alchemical works with this scientific experimentation, technically to "penetrate the recess of nature and to know how she works in her hiding places." Waldman's thinking on this can be traced directly to the real-life Humphry Davy, and, in a nonlinear way, to Paracelsus.

But while in Ingolstadt, Paracelsus was not able to bring someone back to life. Instead, during his days in the town, he did cause a paralyzed woman whose condition was said to be irreversible to walk again. No one knew how he did this, but some said it was brought about by his medicinal lotion called Azoth of the Red Lion. This "universal medicine" contained mercury, which even then was known to possess curative powers. There are accounts of the girl's rising from her bed and waltzing gleefully into the next room, where her parents sat reading.

PARACELSUS WAS NOT THE only person to think of a man-made creature, nor was Mary Shelley the first to write about it. History, lore, and religion have numerous tales of mystical men who

The process of alchemy was at the highest point during Mary Shelley's lifetime. Alchemists were not only trying to turn base metals into gold, but to find the key to immortality. As the illustration shows, it was not only scientists and alchemists who were involved in the practice, but monks secluded in their monasteries. For them, finding the key to immortality was not going against God; rather, it was a way to understand Him and His doings.

created beings large and small—either to help in times of trouble or simply because their physical labors had become too burdensome. Most of these belong to the Jewish tradition and involve a creature called a golem. It was in sixteenth-century Prague that the most renowned golem of all came into existence. It was there that Rabbi Loew removed some soil from the earth and made himself an assistant.

Rabbi Loew knew the city's Jewish population wanted help, but also that his own household needed assistance with the daily chores. As a rabbi, he was aware of the incantations required for golem-making. He didn't need putrid sperm or horse manure like Paracelsus. Rather, a lump of clay, long-forgotten prayers recited in a specific rotation, and the quiet whispering of God's secret name would do the trick. As such, Rabbi Loew and two of his relations dug out clay, which was then shaped into the form of a small man. They stretched out the clay figure on a large table situated in the middle of a room and took turns walking around it seven times, while rhythmically chanting their prayers and whispering to God. Soon, the clay creature began to glow from within, and as more prayers were uttered the creature grew in size. Eventually it opened its eyes and, startled by being brought to life, gazed at them, as if intent on pleasing them.

But as it continued to grow, Rabbi Loew was afraid to let it out of his sight. In addition to increasing in size, the golem also got smarter and smarter and learned to disobey orders, while also overstepping its boundaries. Not long thereafter, it was decided the creature should be returned to the clay. Three men straddled the golem and roped it tightly. Once again the people circled the golem, but this time they did so in the opposite direction, while again reciting prayers that would take away the life force they had given the creature. They also recalled the secret name of God. When this was done, the creature lost its life force and sank back onto itself.

The golem tales were common all over Europe. Jacob Grimm, half of the famous team the brothers Grimm, in his 1808 book *Journal for Hermits* wrote, "The Polish Jews, after having spoken certain prayers and observed certain Fast days, make the figure of a man out of clay or lime, which must come to life after they have pronounced the wonderworkings *Shem-ham—phorasheh*. This figure cannot speak, but it understands what one says and commands it to do. They call it Golem and use it as a servant to do all sorts of housework, only he may never leave the house. On his forehead the word *Aemaeth* (Truth, God) is written, but his weight increases from day to day, and he easily becomes taller and stronger than all the other members of the household, however small he might have been in the beginning. Becoming afraid of him, they therefore erase the first letter so that nothing remains but *Maeth* (he is dead), whereupon he collapses into clay."

Given everything readers of *Frankenstein* knew about alchemy and creating a man, they were left to wonder how much Victor Frankenstein truly knew about man-making when he referred to Paracelsus and Simon Magus. And how much did Mary Shelley know about them when she referred to them? Where had she learned alchemy? It wasn't a matter of where she learned of them but of who had taught her. And that man was a young poet who entered her life early on and, as if she were much like alchemical clay, reshaped the direction of her days. He was Percy Bysshe Shelley.

In 1812 Percy Shelley wrote a letter to William Godwin in which he described not only his movements and current situations but also what he'd been reading. In the letter, he said he had read all of the books about Albertus Magnus and Paracelsus. This admission, which sounded suspiciously like the one Victor Frankenstein makes to his own father, began on a correspondence between Shelley and Godwin, which had developed some months earlier. Shelley had learned that the reformer was still alive and not dead, as he had come to believe, thus had quickly drafted a letter full of youthful enthusiasm, praise, and the proper dose of flattery. The letter worked its magic on Godwin; the reformer quickly replied.

Godwin was used to receiving letters like this from young men who had been inspired by his writing and wanted confirmation

Percy Shelley. This portrait was painted in Rome in 1819, by Amelia Curran. It is the most well-known portrait of the poet, and Mary Shelley went to great lengths to get it from the artist following Percy Shelley's death. Luckily for Mary Shelley, Amelia Curran still had it in her possession—believing it was not a good likeness of the poet, Curran was about to burn it when word from Mary Shelley reached her.

they were pursuing the right approach themselves. In 1803, he had received such a letter from a young man named H. Chatterton. Only twenty-one years old, Chatterton's words were oddly similar to ones Mary Shelley would write years later, and they also echo the ones Victor Frankenstein mutters on pages printed over a decade later.

Chatterton wrote:

> *From my boyish days I was propelled of a sentiment . . . that I was born to promote the diffusion of knowledge. It were absurd to prove to you how much the increase of happiness depends on the progress of myth, and truth on the invention of the adoption of a philosophical knowledge. I have long had this object in view. But what an immense field of science must a man travel . . . It were useless, as well as ostentatious to describe my exertions & especially as their effect in knowledge seems disproportionally small . . . It was only by the faint glimmer of my midnight lamp, in solitude and science, that I could recover my mind from my agitation in which the scenes of the day had thrown it; and that my worn imagination permitted a virtuous and rational state of society inflicted in some measure my own exertions.*

There are no indications that Godwin replied to this letter— though he eagerly replied to the one from Shelley.

SHELLEY'S ENTHUSIASM FOR LIFE'S MORE UNORTHODOX SUBJECTS HAD always been contagious, going so far as to engulf those that surrounded him. In some respects, these enthusiasms caused him to be expelled from Oxford University not long before he came to know William Godwin.

Most of what is known of the six months Shelley spent at Oxford comes from Thomas Jefferson Hogg, a young man he met and befriended soon after he arrived. Though Hogg is believed to have been bright in his own right, and in later life became a lawyer, he is best known for his association with Percy Shelley. He tried to make the most of that friendship, particularly with his book *Shelley at Oxford*, in which he anointed himself the expert on Shelley's schoolboy days.

The book stresses many of the characteristics most people associate with Shelley, especially the beautiful physical traits that seem to elevate the poet from the mere mortal to the angelic.

The initial meeting occurred one evening during the school's scheduled meal. As Hogg sat at his assigned table, he realized he had been joined by a young freshman who was so tall and gawky he gave off an aura of fragility. He wore expensive clothes, though he seemed not to mind them, and there was a gentility about him, though "his gestures were abrupt, and sometimes violent." His long hair crowned a face that was pale, small, and delicate like a girl's. He ran his hands through his hair regularly yet unconsciously. Hogg was captivated and could not help but stare at the young man who had entered his life.

During their initial conversation, Hogg learned that Shelley had a peculiar fascination with science and chemistry, most especially electricity. "What a mighty instrument would electricity be in the hands of him who knew how to wield it, in what manner to direct its omnipotent energies, and we may command an indefinite quality of the fluid," Shelley cried out to a stunned Hogg.

Shelley also spoke about the powers of thunder and lightning, and of the possibilities for man should he be able to "guide it."

Shelley was enthralled by overpowering nature: "How many se-crets of nature would such stupendous force unlock," he said.

He invited Hogg to continue this one-sided conversation in his rooms, not only so that he could expand on his scientific ideas, but also so he could show off one of his most prized possessions: a galvanic battery.

At Eton a few years earlier, he had refused to take part in sports and the same studies as his peers. Instead spent his time on "strange studies," inundating those who would hear him with stories of "fairy land, and apparitions, and spirits, and haunted grounds." He launched fire balloons toward the sky and barricaded himself in his room reading texts on chemistry and demonology. His room often smelled of "strange and fiery liquids" that he kept bubbling on the tables. He also started to experiment on himself with a small galvanic battery he had purchased before entering the in-stitution. These experiments occurred in the middle of the night, while the rest of his classmates slept. Though Shelley left behind few notes about these experiments, his tutor, a Mr. Bethell, who had once been subjected to the effects of the battery, kept an ac-count of what happened.

As Edward Dowden, who researched and wrote a detailed book about Shelley and his scientific pursuits, and T. J. Hogg reported, while on a nightly round Mr. Bethell heard peculiar noises coming from behind Shelley's locked doors. Curious, he became convinced that Shelley was engaging in "nefarious scientific pursuits," which of course he intended to put a halt to. He marched into the room, where Shelley was engulfed in a leaping "blue-flame." Stunned, Bethell asked what he was doing, and Shelley replied, "I am raising the devil."

On hearing this, the tutor approached the galvanic battery and placed his hands above it. He received a nasty electrical discharge that sent him flying across the room. What little bond existed between tutor and pupil was seriously frayed that night.

ALTHOUGH GIOVANNI ALDINI HAD PREVIOUSLY PERFORMED THE most comprehensive galvanic experiment on a human corpse in London, the introduction of galvanic electricity to an English audience had occurred earlier, by the so-called medico-electrician Tiberius Cavallo. It was Cavallo who had brought the work of Galvani to the forefront in England, in 1793, when he read two extended and detailed letters he had received from none other than Alessandro Volta, who was detailing Galvani's experiments on frogs. Cavallo also read those letters at the Royal Society, making them available to scientists in the not-so-distant future.

Since the late 1770s, Cavallo had been dabbling with electricity on a medicinal level, even constructing those instruments he needed for his purposes. This had made him one of the most sought-after and famous natural philosophers in all of London. Aside from his own abilities as an inventor and medical therapist, he was also an excellent letter writer, a trait that led him to begin and keep up friendships and professional correspondences with a wide variety of people across Europe and beyond, people who kept him abreast of the latest inventions and innovations in the fields. One person he corresponded with was Dr. James Lind of Windsor, who would later become one of Percy Shelley's mentors at Eton.

Lind was interested in using galvanic electricity on diseased bodies. Through his many letters, he was kind enough to provide Cavallo with stories and anecdotes about his experiments, the ins and outs of how he used the galvanic battery on his patients. He also let Cavallo know about the results from experiments being conducted by other scientists across Europe; Cavallo did the same, providing Lind with information he received and discovered.

These interchanges between Cavallo and his friends gave him what he needed to complete a major work, *A Complete Treatise on Electricity, in Theory and Practice, with Original Experiments*. It was remarkable on many levels, not only because it explained animal electricity at length, but also because it detailed, in sometimes graphic prose, the experiments some doctors were performing not on dead bodies, but on themselves. Of particular interest was Dr. Munro, who, on one occasion, "applied a blunt probe of zinc to the Sephum Norium, and repeatedly touched it with a crown-piece of silver applied to the tongue, and thereby produced the appearance of a firefly, [and] several drops of blood fell from the nostrils."

But throughout the years Cavallo's most noteworthy and prolific letter writer continued to be James Lind. In a letter written on August 2, 1792, Cavallo piped, "P.S. Have you made any dead frogs jump up like living ones?"

After a fulfilling life, notably as the physician to King George III, Lind settled into a cottage in Windsor with his wife. In Windsor, Lind developed a reputation for "tricks, conundrums, and queer things." Madame D'Arblay, a Windsor resident, suggested that neighbors were afraid of him, of "his trying experiments with their constitution . . . they thought him a better conjurer than physician." His laboratory was a cornucopia of galvanic batteries, metal probes, surgical instruments, dead frogs, scalpels, bubbling vials, gases, and poisons, a chaotic environment that strangely enough would be duplicated by a pupil in his rooms at Oxford University.

When T. J. Hogg met Percy Shelley, he was perhaps hoping for a friendship that would blunt the loneliness that had settled over him once he had entered Oxford. True, he had always wanted to attend "that land of promise." But a certain measure of melancholy had seized him when he arrived. He desperately needed company; Shelley quickly provided that. As such, he decided to visit the "young chemist's" rooms, where he discovered a strange odor emanating from the various gases bubbling forth in the vials resting on the table.

The rooms had been cleaned before the new occupants arrived, but Shelley's resembled a disastrous laboratory. Everything was in disarray, from personal belongings to "philosophical instruments," as "if the young chemist, in order to analyze the mystery of creation, had endeavored first to re-create the primeval chaos . . . An electrical machine, an air-pump, the galvanic through, a solar microscope, and large glass jars and receivers were conspicuous amidst the mass of matter." Shelley quickly walked back and forth about the room, and Hogg learned that a combination of ether and some other powerful fluid had spilled out of the vials and onto the floor, igniting and leaving dark smudges on the carpet.

Shelley pointed a finger toward the electrical apparatus on the table, wishing to show his new friend how the machinery worked. Ever so eager to please, Hogg began to fiddle with the equipment, at which point he began to "work the machine until it was filled with fluid," while Shelley turned "the handle very rapidly, so that the fierce, crackling sparks flew forth." This process became so intense, Shelley's "long wild locks bristled and stood on end."

Hogg recalled that Shelley was jumping back and forth across the room, from idea to idea. The process became so wild, Hogg came to believe that Shelley would "set the college on fire, or that he would blind, maim, or kill himself by the explosion of combustibles." Hogg mused that poison would likely be the end of Shelley. He mixed and combined it in an erratic manner without regarding what dishes, cups, and bowls he was using or bothering to find out if the combinations were even plausible. When offered something to eat or drink Hogg habitually examined the cup or dish to make sure nothing unusual was in it; if he didn't see anything peculiar, he pressed his nose to it in an effort to smell it, though even then he could not be sure.

Hogg and Shelley talked well into the night, the discussion turning from the mysteries of lightning and thunder to the details of the electrical kites Shelley had constructed while a young boy at home. He also told Hogg of the love he had possessed for the macabre, which still fascinated him, and the tales of the spirit world he so loved: ghosts, goblins, and water nymphs.

SHELLEY'S FASCINATION WITH THE LAND OF THE DEAD WOULD PLAGUE him throughout his entire life. In a letter written to Hogg, undated but placed by experts to June 27, 1811, Shelley wrote, "I have been thinking of Death and Heaven for days. Where is the latter? Shall we set off? Is there a future life? Whom should we injure by departing? Should we not benefit some?"

From his childhood days, Shelley suffered from debilitating waking dreams which further enhanced the notion of an existing otherworldly life. They occurred so frequently he often took laudanum to quell them. But instead of helping, this potion caused him to have persistent and raging hallucinations.

Laudanum, or *tinctura thebaica*, is a derivative of opium, and it became so fashionable during Shelley's time, the Victorians were notorious for chugging it down in large quantities to relieve anything that ailed them, from gout, to the stress of migraines, to hair loss, even administering it to teething children to ease the pains in their gums. In the latter cases, some of the wet nurses were found to be a little zealous with their ministrations, so much so that some children in their care died of opium poisoning. Laudanum was even sold under such attractive names as Mother Bailey's Quieting Syrup and Godfrey's Cordial. It did not seem dangerous, merely a step removed from the soothing calmness achieved by milk and honey.

Many of the alchemists, including Paracelsus, dabbled with opium. When Paracelsus was conducting experiments dealing specifically with *Papaver somniferum*, the plant from which the opium was derived, he discovered that opium's soporific abilities were a great relief to suffering patients. He continued to investigate the drug's properties and was so astounded with the results, he derived the name *laudanum* from the Latin word *laudare*, "to praise." Unfortunately, Paracelsus did not realize that opium, in all of its forms, was also addictive, causing a whole score of problems for its users.

Percy Shelley, who was "highly sensitive to pain, easily excited, and subject to paroxysms of passions," was one of the many addicted to laudanum. And unfortunately, it did nothing to prevent his waking dreams. His sleep was so disturbed, he often sleepwalked in the middle of the night, his hair wild and disheveled, his face as pale as one of those ghosts he often spoke of. He and his cousin Thomas Medwin lived in the same dormitory, and one night Shelley arose and dragged himself into Medwin's room. Open-eyed but asleep, Shelley walked toward an open window. Medwin jumped from his bed and took Shelley by the arm.

"He was excessively agitated," Medwin later recalled, "and after leading him back with some difficulty to his couch, I sat by him for some time, a witness to the severe crethism of his nerves which the sudden shock produced."

Despite the negative effects the drug could cause, Shelley still hoped that laudanum would eventually give him a reprieve from his sleepless nights and feverish nightmares, which were often bothersome but could turn dangerous. But it would not be very long before he met someone, a kindred spirit, who would understand perfectly. In fact, according to literary history, it was during one of her own waking dreams that the young Mary Godwin first saw "the pale student of unhallowed arts kneeling beside the thing he had put together."

IN THE SUMMER OF 1814, A MEETING BETWEEN PERCY BYSSHE SHELLEY and Mary Godwin took place. Soon after Shelley had sent William Godwin the letter that started their friendship, Shelley began calling on the Godwins at their Skinner Street home. It is possible that Shelley and Mary Godwin met in 1812, when she returned from a period of time abroad. But if so, neither one ever mentioned it.

The situation in the Godwins' household had become not only unhappy but also downright intolerable, thanks in part to the new Mrs. Godwin's attitude toward her stepdaughter. As such, William Godwin removed the source of those disagreements: Mary. The reason given was that Mary's health had become so poor the only way to improve her constitution was a trip abroad, but that was a cover-up for the real story.

Mary was sent to Dundee, a small harbor city in Scotland. Prior to the trip, William Godwin had written to William Thomas Baxter, a man he had met only once before. Baxter's daughter, Margaret, was also married to one of Godwin's friends, David Booth, further cementing their connection. The Baxters agreed to take in Mary Godwin for a short visit so that she could recuperate from an ailment William Godwin had only described in the briefest details. Mary departed on June 7, 1812, on the *Osnaburg*. While she was staying with the Baxters, she struck up a close friendship with the two youngest daughters in the family, Christina and Isabel, becoming especially close to Isabel. At the time of her visit, Dundee was well known for several prestigious industries, such as the manufacturing of jute and fabric. More than anything, Dundee was renowned for its whaling industry and the history associated with it.

Dundee reeked of boiling whale blubber and had a population of between thirty and forty thousand inhabitants at the time of Mary's visit. Most of them were involved in some way or another in the whaling industry, whether as mariners on the many vessels that left the city's small enclosed harbor for the arctic seas, or on land, where the "whaling boiling yards" were located.

Each whaling company had one of these yards, where the whale oil and blubber were eventually transformed into the products needed. Even the women were employed in the field, "to clean the whale bone of its flesh and other impurities."All of this bustling and brisk business in whale products created an awful smell. As late as 1825, an article that appeared in the *Dundee Advertiser* mentioned the "most disagreeable suspicious sort of smell [that] has accosted the olfactory nerves of the inhabitants of this town." Still, the town relied on the whaling business, and the smell was part of it.

Mary Godwin must have been aware of the smell, but more important, she must have heard the many tales that abounded in Dundee. Most of them were associated with the frigid arctic seas, with the mariners and the vessels lost in the watery abyss, with those men who never returned. One such tale was that of Captain Adamson, who manned the ship *Advice.* He had been captured during the Battle of Camperdown in 1797 and brought onto a Dutch vessel, which then sank. Captain Adamson was lost for several days in the vastness of the arctic waters, alone with his thoughts and fears, before he was found anew and returned to safety. Such tales of bravery and despair would no doubt have reminded Mary of Coleridge's *Rime of the Ancient Mariner.* In it, the mariner had also been lost at sea for a time, lost companions, feared the outcome, but was eventually found, and later told his story again and again to those who would hear of it.

But in 1814, when Percy Shelley returned for a visit to Skinner Street, Mary was there, having just come back from another trip to Scotland. Shelley's eyes must have settled on her and for the first time he must have seen the young woman she had become. Edward Dowden described Mary at that age as "a girl in her seventeenth year, with shapely golden head, a face very pale and pure, great forehead, earnest hazel eyes, and an expression at once of sensibility and firmness about her delicately curved lips."

Aside from fancying her physical features, Shelley found something far more important and appealing: Mary was the daughter of William Godwin, the illustrious mentor whose work Shelley idolized, and her mother was Mary Wollstonecraft, the first feminist, whose work *A Vindication of the Rights of Woman* had had such an impact. And Shelley came to believe that Mary had her parents' intellectual gifts, although there was no indication of this yet.

"The daughter of Godwin and Mary Wollstonecraft had gifts of heart and minds such as Shelley had never hitherto known in a woman," Dowden wrote. "From her father she had inherited clearness and precision of intellect, firmness of will, and a certain quietude of manner . . . under this quiet bearing lay her mother's sensitivity and ardour, with an imaginative power which quickened and widened her sympathies."

Percy Shelley was enthralled, and so was Mary. But the situation was also not that unusual. Although his wild looks and tendencies to enjoy scientific discourses and tales of the macabre had made him unpopular with his schoolmates and earned him the nickname of "Mad Shelley," those same traits were very appealing to women.

By the time Shelley became a constant guest at Skinner Street, three young ladies were living in the house. There was quiet and reserved Fanny, the one people spoke of the least and the one who was the most homely. She was older than the other two and obeyed the house rules; some said she did so because she felt like an outsider, given that neither Mr. Godwin nor Mrs. Godwin was her natural parent. Mary was the somewhat pretty middle girl, who was quiet but not too quiet, and who leaned toward reflection and introspection. She was the inheritor of her father and mother's legacies but was far too attached to her father, something her stepmother had tried, without success, to end. And there was the youngest, Jane (Claire) Clairmont, the most beautiful of the three, dark and alluring, bold in her looks and manners. All three, it was suspected, formed a crush on Shelley, but only Mary had the mental capabilities and legacy he was attracted to.

Mary often made it a point to visit her late mother's grave site. St. Pancras Cemetery's northward location offered her solace on two levels: it got her away from her stepmother and afforded her a spiritual friendship with her dead mother. She would sit on that grave site and read most of Mary Wollstonecraft's published and unpublished works, as well as other manuscripts she removed from her father's library.

Like other cemeteries across London and its suburbs, St. Pancras had its share of resurrectionists and "sack-'em-up men" disturbing graves for profit; Mary Wollstonecraft's site, though, had not been touched. Jane Clairmont often went with Mary on her strolls, not because she too wanted to read in the cemetery, but because young girls should not go out alone. After Shelley started coming around, he joined their walks toward the cemetery, this time with Jane acting as a young chaperone.

Shelley must have liked the gloomy mood of the cemetery and the adjacent church. He likely told the girls about his obsession with the "marvelous stories of fairy-lands, and apparitions, and spirits, and haunted grounds." St. Pancras, with its old history, its towering weeping willows, and its dank river nearby, could easily bring back days from the past. But St. Pancras appealed to Shelley because while there, beneath the gnarled tree sweeping above the tomb of Mary Wollstonecraft, he realized he was in love with Mary Godwin and told her so. As it happened, an exuberant Mary said she felt the same. Only one thing stood in the way of their happiness: Percy Shelley was married, and he was the father of a toddler named Ianthe.

PERCY SHELLEY AND T. J. HOGG MAY HAVE DISCUSSED THE WONDERS of science and the pursuit of knowledge in their rooms at Oxford.

But invariably their conversations turned to women—they were, after all, young and at the brink of adulthood. Shelley seemed to place great emphasis on the qualities a woman had to possess in order to capture his senses. According to Hogg, Shelley would only be moved by a female if she possessed "absolute perfection."

Edward Dowden agreed: "It is certain that at this time the qualities in women which most kindled [Shelley's] imagination were not beauty, or sweetness and gentleness, but intellectual strength and passionate ardour of heart."

Some said Shelley's wife, Harriet Westbrook, had none of those qualities, though nothing was outright wrong with her. She was a beautiful girl of sixteen, who was very petite and "slightly and delicately formed." She moved quickly on her feet and dressed in so simple a manner as to match her speech and the tone of her conversations, and "her laugh was spontenous, hearty, and jouyous."

But people said she was not as intellectual as Shelley wanted, and he did not care much about beauty and fashion, so why did he marry her? The answer is that Shelley enjoyed playing a rescuer. Before marrying, Harriet lived at home with her father and an un-married older sister, who had stepped into the role of mother early on. Her mother still lived, though she was a "seemingly incapable person." Harriet, in all the wisdom of youth, had found this situation unbearable; when she met the dashing poet she quickly fell in love. In turn, after seeing her situation, Shelley promised to care for her.

But that promise was seemingly made in the heat of the moment, for in a letter he wrote to Hogg, dated on or around August 3, 1811, he declared, "In consequence . . . she has thrown herself upon my protection . . . Gratitude and admiration all demand that

I should love her for ever . . . She had become violently attached to me, and feared that I should not return her attachment."

He proposed, and the two of them rode off to Edinburgh, where they got married. They left behind them a disgruntled father and family.

It did not take long for the Shelleys to realize that Edinburgh would not agree with them, particularly not with Percy. He could not tolerate the persistent rains of autumn and the city's general darkness. After only five weeks, he had no problem convincing his wife to move to York: "she was docile, and submitted her mind to such influences as were brought to play upon it." Even worse, "strength of intellect and strength of character were lacking in her."

Some agreed that this was a particularly terrible flaw in her character, not so much because Harriet submitted herself to Shelley's will and moods, but because she could not keep up with his discussions on philosophy, science, and the occult. And she did not appear to have a mind of her own. She even continued to rely on her older sister for advice well into the marriage, a habit Shelley deplored so much he came to believe he had married both sisters, not just the younger one.

While they were living in York, Shelley's friendship with Hogg was challenged and redefined. Shelley left Harriet under Hogg's protection while he went away to London. When he returned, he noticed his wife was despondent and quiet. He urged her to tell him what had happened, and she revealed to Shelley that Hogg had made a pass at her, declaring his love for her and hoping she loved him back. When confronted, Hogg admitted his indiscretions, and the two didn't speak for many painful months. Again, Shelley decided he needed a change of scenery and wanted to move—something that would occur again and again. He left the choice of where to his wife, and to her sister Eliza, who was now a constant household companion.

Things got worse when Harriet became pregnant and gave birth to Ianthe. Like many women, Harriet transformed with motherhood, shifting her priorities to her child. She matured physically and mentally, though she still relied mightily on her sister. This bond only got stronger with the birth of the baby, and Percy Shelley felt left out. By this time, Shelley had reconnected with Hogg, and on March 16, 1814, he wrote to his friend: "Eliza is still with us . . . I am now little inclined to contest this part. I certainly hate her with all my heart and soul. It is a sight which awakens an inexpressible sensation of disgust and horror, to see her caress my poor little Ianthe."

In 1894, decades after Percy Shelley became romantically involved with Mary Godwin, Mark Twain published a long article titled "In Defence of Harriet Shelley." By this time, Harriet, who had died in 1816, had long since been forgotten. Regardless, there were still questions about her days with Shelley, most specifically what had prompted her husband to land in Mary Godwin's arms. Mark Twain, like many others before him, didn't really know much about Harriet Shelley's life. Still, he was able to conclude that "rumour, gossip, conjectures, insinuation, and innuendo," perpetrated by Shelley's friends and supporters and Mary's family and friends, had contributed to marking Harriet as a lowly character.

Later, after Percy Shelley and Mary Godwin eloped, a smear campaign against Harriet Westbrook Shelley broke out. She was

portrayed in many circles as a cold and indifferent wife who did not
fulfill Shelley's needs. She was described as uneducated, mentally incapable of keeping up with her husband's intellectual exercises, and,
worse still, as the daughter of a tavern owner (among other establish-
ments). People even began to say she had likely been a drunk.

This portrait came as a surprise to those who actually knew
Harriet. The real woman was quite different. She had been well
educated and attended the same school as Shelley's sister (who
came from money). In that respect, she had attended better schools
than Mary Godwin, whose education and knowledge were derived
not from formal schooling, but from her own efforts and desire to
learn, and the steps she had taken to gain that knowledge.

It also was not true that she had been indifferent to Shelley's
needs. Thomas Love Peacock, a friend of Shelley's, wrote in *Memoirs of Percy Shelley*, published in 1858, that Harriet "was fond of her
husband, and accommodated herself in every way to his tastes. If
they mixed in society, she adorned it; if they lived in retirement,
she was satisfied; if they travelled, she enjoyed the change of scene."

In his 1857 book about Shelley, Edward Dowden portrayed
Mary as a calm and particularly reasonable young woman, even
though her bouts of anger and sour moods were well known and
documented. Others also took issue with Mary's features and how
she was portrayed physically. It was said that she was an exceptionally beautiful young woman Shelley could not resist, but according
to Shelley's cousin, Thomas Medwin, that was not entirely true.
He said: "It could not have been her personal charms that captivated him, for to judge her [Mary] in 1820 . . . she could not have
been handsome, or even what may be denominated pretty."

If Shelley had sought beauty, he had already had that. Harriet's physical beauty was startling, as were her many personal charms. Those who met her were immediately put at ease by her presence, something Mary often failed to do.

Those who came to learn of Shelley's subsequent romantic adventures knew very well why his wife had been disposed of and *that* particular mistress gained. Even Harriet knew why she had been set aside. When asked this by Thomas Love Peacock, she replied, "Nothing, but that her name was Mary, and not only Mary, but Mary Wollstonecraft." Mary Wollstonecraft Godwin, at that. Harriet might have felt slighted at that moment, and may even have been angry, but her response was not too far from the truth.

How does one go on about explaining falling in love with a name, or the notion of what that name actually means? How does one rationalize that decision or impulse? In Shelley's case, he convinced himself, Dowden agreed, that "he had married a woman who . . . had never truly loved him, who loved only his fortune and his rank, and proved her selfishness by deserting him in his misery." By shifting the blame toward Harriet, Shelley freed himself of fault. He was available to pursue his own happiness, which he thought he'd found with Mary. He had had numerous amorous escapades, was flexible in his romantic entanglements, and made it seem as if he had committed a pardonable act. Mark Twain suggested that Shelley "had done something which in the case of other men is called a grave crime," but said that "in his case it is not that, because he does not think as other men do about those things."

Harriet also faced rumors that she had been unfaithful and that the child she was carrying (her second) might not have been Shelley's. People figured the Godwins were to blame for this gossip—Mrs. Godwin was known for her fondness of gossip and flair for the dramatic. And William Godwin could have started such a rumor to save what was left of his daughter's reputation or to solidify his relationship with Shelley, who had already extended financial support and would continue to do so in the future. Callous as it sounded, those closest to him were aware that this behavior was not above him.

"As respected his own purposes, Godwin was one of the most heartless, the most callous of men," Francis Place wrote. "He was perfectly regardless of the mischief he might bring upon anyone and quite as regardless of the feelings of others. When his own ends could be best and must be promptly answered by inflicting unhappiness on them, these matters annoyed him so little that I have sometimes doubted whether they did not even afford him satisfaction, when they fell upon those who had not readily conformed to his wishes."

MARY GODWIN KNEW SHELLEY WAS MARRIED WHEN THEY MET. BUT Shelley had told her the marriage was dead, though no one else had been aware of the troubles he was facing. He described for Mary the cruel situation he found himself in and how desperate his soul was. He most likely told Mary that Harriet had been unfaithful. Whether he truly believed that is uncertain, but how was Mary to react to that? She was a young woman experiencing her first full-blown love affair, believing the object of her desire was caught in the middle of an awful predicament.

After several conversations with Mary and Percy Shelley, William Godwin believed their passion and desire for each other had been diffused, but that was not true at all.

Harriet had also come to believe that her husband's zest for the young Mary would wane. In her mind, she thought that Mary was the one "who had ensnared Shelley by her witchcraft, by her sentimental raptures at Mary Wollstonecraft's grave; by her avails of love; she alone had done this wrong."

Given their situation, Percy Shelley and Mary Godwin decided to elope to the continent. On July 28, 1814, cloaked in a long dark dress that matched the night, Mary Godwin and Percy Shelley met at dawn to rush to Dover, and from there they intended to go to Switzerland.

Anchored by their sides was none other than Jane Clairmont.

Chapter 5

ELOPING TO THE MAINLAND

The castled crag of Drachenfels

Frowns o'er the wide and winding Rhine,

Whose breast of water broadly swells

Between the banks which bear the vine,

And hills all rich with blossom'd trees,

And fields which promise corn and wine,

And scatter'd cities crowning these;

Whose far white along them shine . . .

LORD BYRON, *CHILDE HAROLD'S PILGRIMAGE*

PERCY SHELLEY, MARY GODWIN, AND JANE CLAIRMONT fled Skinner Street during a blistering hot night, a heat wave that would stay strong as they traveled across the channel, accompanying them well into the Continent. Both girls were about to get an adventure and they eagerly hurried toward the chaise Shelley had provided. Shelley had been unsure if Mary would actually show up, but as the new day dawned, there she was.

Was he surprised to see Jane tagging along with Mary? After all, this was supposed to be a romantic elopement, and Jane was a third wheel. Why would Jane accompany Mary and why would Mary agree to it? When news of their escape broke out, some wondered if Jane had also developed a crush on Shelley and wanted to intrude on her stepsister's happiness. Many others suggested that Jane was with them because she was fluent in French, and they would be traveling through France and part of Switzerland. (But of course Shelley spoke some French and could have done well without her.)

Still others speculated that Jane had also aroused in Shelley the hero-rescuer role he was so fond of playing. He had done so when he met and married Harriet, and he was doing it again with Mary by taking her away from circumstances she despised. So why should it be any different with Jane?

Others recognized that Shelley's moral standards were flexible and were inclined to believe he had run off with the two girls. Shelley had often spoken about starting a commune, and this idea, along with Shelley's unconventionality, was further bolstered when later he dared to ask his wife, Harriet, to join them. He sought her company not as his wife but as his friend. It's possible he wanted Harriet to bring money with her—or he enjoyed adding fuel to the malicious and widespread rumors that had already started. But Harriet declined the invitation.

William Godwin learned of their escape that same morning. He found a letter on his dresser that told him his daughter had escaped behind his back. In his younger years Godwin's own views and ideas about love, relationships, and marriage had been as loose as Shelley's, but he was not as open to his daughter and his protégé engaging in such potentially scandalous activities.

In writing to his friend John Taylor sometime later, he placed the blame on Shelley, though he felt betrayed by all the participants: "I had the utmost confidence in him; I knew him susceptible of the noblest sentiments; he was a married man who had lived happily with his wife for three years . . . On Sunday June 26, he accompanied Mary, and her sister Jane Clairmont, to the tomb of Mary's mother, one mile distant from London; and there it seems, the impious idea first occurred to him of seducing her, playing the traitor to me and deserting his wife."

Godwin was wrong, of course. Though Shelley may have made his feelings clear to Mary, all reports indicated that she was the one who made the first move. Shelley *had* spoken to Godwin earlier about his feelings for Mary, almost as if he was asking permission to begin a relationship with her, but Godwin had as much as said no.

"I expostulated with him with all the energy of which I was master, and with so much effect that for the moment he promised to give up his licentious love," Godwin went on in his letter to

Taylor. "I appealed all my diligence to waken up a sense of honour and natural affection in the mind of Mary, and I seemed to have succeeded. They both deceived me."

After this conversation between her father and Shelley, Mary decided, at least for a short time, to blunt Shelley's advances. Ultimately she decided to run away with him, perhaps choosing that as a way of avoiding her father's anger. One story at the time, told by Mrs. Godwin, suggested that Shelley attempted suicide by ingesting a large quantity of laudanum. Mary found him and revived him, and he resumed his seduction of her. This time he won her affection.

During their journey across the channel from England to France, they were caught in a severe thunderstorm that churned up the seas. Crossing torrid waters had always caused Mary to become violently nauseous and this time was no different. But scholars have speculated that Mary might have been suffering the initial pangs of morning sickness, due to early pregnancy.

The Godwins always claimed that Mary and Shelley's sexual relationship began after their elopement (when Percy had officially, if not legally, left his wife), but Mary and Percy's works and diaries indicate that the affair had become intimate weeks earlier, in, of all places, St. Pancras Churchyard. Surrounded by tall, old weeping willows and ancient tombstones—including Mary Wollstonecraft's—the churchyard gave them the privacy they sought, away from any watchful eyes. And if Jane had chaperoned on those occasions, as she always did, she could have stood watch to make certain no one was around while they went about their business, and made herself scarce when asked. She said she was most often sent away once they reached the cemetery.

"They always sent me to walk some distance away," Jane later wrote, "alleging that they wished to talk philosophical subjects and that I did not like or know anything about those subjects."

As the boat bobbed from side to side and the three of them clung to one another, Mary's nausea got worse. And when they disembarked, Mary seemed to become deeply frustrated, her snide remarks including insults to the locals. She compared them unfavorably to the English by scrutinizing their clothing, manners, looks, and general attitudes, declaring, while in Calais, that "unfortunately the manners are not English." Her behavior could have been due to the storm's severity, her physical discomfort, or Jane's presence, but whatever the reason, it was noticeable to everyone.

France was experiencing the same vicious heat wave, but at least in Paris they were able to stroll along the wide boulevards and lush green gardens like tourists, though even there they waited for a sum of money that would set them "free from a kind of imprisonment which [they] found very irksome." Percy knew that when he abandoned his wife, he would not be able to count on his family's financial support for this new adventure. Mary, as well as Jane, had no money of her own to contribute to the party. Shelley continued to be in contact with his mother and sisters, who, not only then but throughout his life, offered the little support they could. Once again, he wrote to them and awaited whatever they could send, however minimal.

Eager to leave the city as soon as the money arrived, they conceived a plan that was not only romantic but silly and delusional: they decided "to walk through France." The idea that three young people—two teenage girls burdened by long frocks, one of them probably pregnant, and a frail-looking man who was no more than a boy himself—could cross an entire country on foot in the middle of summer must have seemed ludicrous even to them. They bought a donkey to carry their supplies and planned to take turns riding him during the journey. But this donkey was skinny and feeble and could not even carry their provisions. The animal's legs buckled when weight was placed upon it. Still, undeterred, they departed.

Mary delighted in the natural views. Despite the people, the heat, the trudging on foot along pebbled roads and through ruined villages, she still clung to the notion that beauty might be found in the natural surroundings they encountered. "A rocky hill rose abruptly on one side, on the top of which stood a ruined citadel with extensive walls and hives," she recalled at one point. "Lower down but beyond, was the cathedral, and the whole formed a scene for a painting."

Given Mary's love of landscapes, it's not surprising that *Frankenstein*, aside from the deep philosophical questions and debates it raised, always managed to astound its readers also with its depiction of the natural world that surrounded its characters. In it, in the loneliness and isolation of the thick Swiss and German forests, the characters always find time to revere their landscape, to bask, while exploring nature, in the idea of something bigger and beyond them, something that, not unlike a forest, was at once as forbidden as it was eager to be penetrated. But Mary, Percy, and Jane did not find beauty in everything they passed; desolate walkways, lonely towns, and barren villages ruined by war seemed to quiet their spirits and sour their moods.

That gloominess was lifted upon crossing the border into Switzerland. Mary wrote, "The scenery of this day's journey was divine, exhibiting piney mountains, barren roads, and spots of verdure surpassing the imagination." True, the landscape was magnificent, but there must have been a different reason as to why Mary felt, at first, so profoundly attracted to the land: this was the country her mother had wished to spend time in. Their primary goal was to make "a journey toward the lake of Uri, and seek in that romantic and interesting country some cottage where [they] might dwell in solitude." Their plans were once again thwarted by money, or the lack of it. They finally accepted that neither of their families was sending anything and they had better return to England.

To do so, they had to pass Lucerne, which was dominated by a lake of the same name, and that is where Mary began collecting local tales. Aside from offering natural beauty, Lucerne was also ripe with old legends and stories. "The summits of several of the mountains that enclose the lake to the south are covered by eternal glaciers," she wrote. "On one of these, opposite the Brunen, they tell the story of a priest and his mistress, who, flying from persecution, inhabited a cottage at the foot of the snows. One winter night an avalanche overwhelmed them, but their plaintive voices are still heard in stormy nights, calling for succor from the peasants."

The fact that she described this short tale not only shows her interest in folklore and the stories of a particular region, but also displays her propensity to use what she had heard in her own work and to mold the tales in order to make them fit her needs.

She used the story in the first edition of the *Frankenstein* tale, where the legend she had heard now read: "I have seen this lake agitated by a tempest, when the wind tore up whirlwinds of water, and gave you an idea of what the water-spout must be on the great ocean, and the waves dash with fury the base of the mountain, where the priest and his mistress were overwhelmed by an avalanche and where their dying voices are still said to be heard amid the pauses of the nightly wind."

As they traveled, and what money they had began to run out, it became obvious that they couldn't afford to take a coach. As it was, water offered the best solution for a return to London, most specifically, the Rhine River.

With their dream of a romantic elopement dashed, Mary's mood darkened even further, and she became almost remorseless in what she said to the people she encountered. As they left Switzerland behind, she could not help making one last jab: "The Swiss appeared to us then, and experience had confirmed our opinion, a people slow of comprehension and of action," she wrote. "But habit has made them unfit for slavery, and they would, I have little doubt, make a brave defense against any invader of their freedom."

On August 28, the Shelley party boarded a boat on the Rhine.

At any other time, traveling along the sublime beauty of the Rhine would have been a magnificent experience. Travel on Europe's major rivers became quite common during the second half of the eighteenth century, among not only the wealthy but also those wishing to move up the ladder of prosperity. Some took such journeys for the simple pleasure of viewing the panoramic landscapes, but others were transporting goods from one end of the country to the other. Still others were taken in not only by the natural beauty that fanned across their eyes, but by the histories they encountered. The Rhine in particular, rushing from the Alps and snaking along its path to the North Sea, crossed boundaries steeped in myths, legends, and fables. All of those were surrounded by landscapes that varied remarkably and could display at once a patchwork of gentle vineyards rising up the slopes of a hill or the ugly turns of the Via Mala, the Evil Way, so called because the extraordinary gorge bearing the same name narrowed at a certain point before plunging malevolently into the river.

Some travelers might have learned the legend that surrounded the Rhine, that of the Lorelei, and journeyed down the river's harrowing waterways in the hopes of debunking it. Lorelei was a German girl who had learned of her lover's unfaithfulness. Soon thereafter, she committed suicide by jumping into the river, where, upon her death, she was immediately turned into a mermaid. From then on she spent her days on a rock near St. Goar, chanting a melodious tune that so enraptured sailors passing by that they eagerly rowed toward her. Unfortunately, the sadness from her voice pulled them too close and they rowed toward their deaths, because the rock she sat upon was located in the deepest and most impenetrable portion of the river. The legend became well known not only in Germany, but also throughout Europe, especially upon the publication of Heinrich Heine's poem of the same name, "The Loreley," which read in part:

> *A song of mysterious power*
> *That lovely maiden sings*
> *The boatman in his small skiff is*
> *Seized by a turbulent love,*
> *No longer he marks where the cliff is*
> *He looks to the mountains above.*
> *I think the waves must fling him*
> *Against the reefs nearby*
> *And that did with her singing*
> *The lovely Loreley*

Nearly two centuries later Sylvia Plath would become entranced with the legend, writing: *"Of your ice-hearted calling / Drunkenness of the great depths / O river, I see drifting / Deep in your flux of silver / Those great goddesses of peace / Stone, stone, ferry me down.*

Mary Godwin seemed chilled by her various traveling companions and thought little of them: "Our companions on this voyage were of the meanest class, smoked prodigiously, and were exceedingly disgusting," she wrote, seemingly repelled. "There were only four passengers besides ourselves, three of these were students of the Strasburg University: Schwitz, a rather handsome, good tempered young man; Hoff, a kind of shapeless animal, with a heavy, ugly, German face; and Schneider, who was nearly an idiot, and in whom his companions were always playing a thousand tricks."

The area they were cruising along had been vividly described by Lord Byron and was now coming to life before their very eyes. Aside from viewing their surroundings, they also spent time reading from a book of Mary Wollstonecraft's they had carried with them. Soon they passed "a ruined tower with its desolate windows [that] stood in the summit of another hill that jutted into the river."

On September 2, they reached the city of Mannheim. As soon as they docked, Jane wrote in her diary: "We arrive at Manheim early in the morning—breakfast there. The town is clean and good. We proceeded towards Mayence with an unfavorable wind. Towards evening the batelier rests just as the wind changes in our favor. Mary and Shelley walk for three hours; they are alone."

Famously, those three hours Mary and Percy spent alone in that particular geographical area have, through the passage of time and more careful reading, given rise to much speculation, because the stopover would have given them time on their own, away from the ever-present Jane, and time to explore their surroundings, most especially those small towns and castles lining the lower banks of the Rhine. One such town was Nieder-Beerbach, on whose summit, barely visible from the water's edge, stood the famed, or infamous, Burg Frankenstein.

"What's in a name?" Mary Shelley wrote years later in a book titled *Rambles in Germany and Italy*. "It applies to things known; to things unknown, a name is often everything: on me it has a powerful effect; and many hours of extreme pleasure have derived their zest from a name."

Following Mary's own words, it could be said that the name of Victor Frankenstein had not come about by mere coincidence, chance, nor in one of her waking dreams, as she always claimed. Rather, she had given her character's name much thought and consideration. But Frankenstein was not a popular name, especially not in England, so where had she heard of it? One theory, originally put forth by the historian Radu Florescu, suggests that during that three-hour walk, she got the inspiration not only for the name, but for the book's basic narrative thread as well.

CENTURIES BEFORE MARY SHELLEY BROUGHT THE NAME INTO THE limelight, the surname of Frankenstein had already been tied to both fact and fiction, most particularly in the German region of the Rhineland. The real Frankenstein family had settled in a formidable castle overlooking the Darmstadt region, where their deeds, famous and infamous, began to be recorded in the annals of history.

The castle itself rose behind unbridgeable mountains, its outline shivering against the gray sky that most often covered the region. In the mid-1400s, the castle was the site of much bloodshed when a member of the family was locked in mortal combat with an enemy of unusual fortitude and cunning, with a deep understanding of psychological warfare. The enemy, intent on overtaking Burg Frankenstein, had successfully overthrown other families in the past. Known for his brutality, Vlad the Impaler and his doings provided, in part, inspiration for another gothic masterpiece: Bram Stoker's *Dracula*.

Within the church located at the bottom of the castle, a brass relief revealed the figure and gory tale of an additional Frankenstein family member, Sir George Frankenstein, a knight who had lived in the burg in the sixteenth century. Sir George's death had thrown the family name from the historical to the legendary, for his mythical battle had been fought not against an enemy of the mortal kind, but of the supernatural: a fire-spewing dragon. It was said that following a fierce battle, Sir George managed to pierce the dragon's heart with his lance, but not before the dragon's tail found an opening in Sir George's armor and inflicted a deadly wound.

Did Percy Shelley and Mary Godwin hear those stories? Certainly, while the boat was docked in the area, those three hours away from Jane might have offered an excuse and opportunity for exploring the surroundings. While the time frame might not have been enough to allow them to hike up to the castle, it still could have given them ample time to visit the adjacent village of Nieder-Beerbach and to talk to its inhabitants. In the shadows of the thick forests, the people might have told them the mythical legends populating their woods and castles, and if this occurred, both, Shelley in particular, with his fascination with the occult and the mystical, would have eagerly listened on. Continuing on that thread,

if they remained long enough to hear about the Frankensteins'
bloody battles against Vlad the Impaler and of Sir George Frankenstein slaying the dragon, then they most certainly would have
heard about the castle's most notorious inhabitant, Johann Konrad
Dippel, a man who, strangely enough, bore a striking similarity
to Victor Frankenstein, and to an extent, to Percy Shelley as well.

Johann Konrad Dippel was related to the Frankensteins not by
blood but, in a sense, by birth. He was born in Burg Frankenstein
on August 10, 1673. Because of that detail, Dippel always felt a
strange affinity to the family, at times going so far as to declare
himself a Frankenstein. In reality he was the son not of nobility
but of a Lutheran clergyman whose intent upon Johann's birth
was to make of him a clergyman, thus continuing the family's
long-standing tradition. Though he actually ended up studying
theology at Giessen as his father had wished, it became painfully
obvious that Johann from a young age possessed doubts about his
father's religious convictions and would not make the perfect clergyman, nor even a mediocre one.

While at Giessen his ideas and individual thoughts were further
tested by the theologian and historian Gottfried Arnold, who was,
during Dippel's attendance, a professor of church history. Unlike his
father, whose religious views were strict, Dippel must have noticed in
Arnold a more flexible idea of church doctrine, divinity, and mysti-
cism. And whether it was Arnold's influence or the freedom he felt
at being away from home, Dippel's own ideas morphed.

Soon he left Giessen for Wittenberg, and later for Strasbourg.
Even though there was no doubt that Dippel was creative and of
unusual intelligence, his personal traits seemed to rub people the
wrong way and eventually were a hindrance. His excitability for the

subjects he spoke of often gave rise to what some considered loud and uncalled-for displays of emotions and passions. He became fond of debating those of opposing views, and those debates could either be verbal tongue-lashings or actual physical duels. One such duel at Strasbourg caused him to flee the city, as sparring against a man with a different opinion, he killed him. Rumors also persisted that he had begun to raid the cemeteries in search of bodies, although there was no proof of that. Again he scuttled away in the middle of the night and returned to Giessen.

When Dippel was at Giessen, the study of alchemy was being undertaken in earnest. During that period, the alchemists were busily trying to find the mythical philosopher's stone. For centuries this had been thought to be a substance that would turn base metals, like lead, into gold, but more importantly, it would prolong life. Dippel had no experience in the art of transmutation, but he felt that one way to gain such experience was to read the works of those who had come before him. One such work was Raymond Lully's *Experimenta*. Dippel must have found in Lully a certain understanding, for his texts were a blend of theology and philosophy, of faith and logic, all of which, Lully believed, when merged together could obliterate the mysteries of the supernatural world and give it more rational meaning.

This divinity-based alchemy made sense for someone like Dippel, a new philosopher and inquirer into the bigger questions of life, someone who had studied theology and was, in addition, the son of a clergyman. It made so much sense, he willingly indebted himself to buy new equipment and build a state-of-the-art alchemical laboratory for himself.

Percy Shelley must have noticed that he and Dippel had a lot in common: while they were each young and inexperienced, they embarked on experiments they weren't familiar with and invariably made a score of mistakes.

In that laboratory, villagers believed Dippel had indeed found the formula that produced gold and was using it to purchase lands and homes for himself. The villagers would have felt it was sacrilegious to use the philosopher's stone for one's gain instead of employing it for the benefit of humanity. Thus, such an act, compounded by his own ignorance of how to actually work the material (he left the vials he was using on the flames for too long), caused his equipment to explode, setting fire to all he owned. In the process he lost not only his physical belongings, but also something far more precious—the secret recipe he had devised to turn metal into gold. He tried to begin anew, to buy new equipment on credit, but he could not recollect what he had used, the dosage, and the steps he had taken. All resulted in failures.

Dippel's foes, those who disliked him but inwardly believed he had managed to find the philosopher's stone, were glad he had lost the recipe because they felt that someone like him did not deserve to possess it to begin with. And if sheer stupidity had caused its loss, so much the better. Dippel also had to worry about the disgruntled clergymen in the area, those who had heard about his doings and were unnerved by him. How dare he fiddle with the mysteries of creation? they asked. How dare he believe himself a god, capable of prolonging life, or even creating it anew? And the villagers saw him as nothing more than the devil's minion, someone whose soul had been sold in exchange for forbidden knowledge. In a short time, he had managed to anger and alienate everyone he knew and a good number of people he didn't even know.

It didn't help that soon after the fire incident, he turned his attention to finding a "universal medicine." He was not the only one seeking this universal cure-all, whether it was a lotion or a balm. Paracelsus had also believed in his Azoth of the Red Lion's ability to aid his patients. Such concoctions were numerous and could involve hundreds of ingredients from the natural world. One such book popular at the time was Robert Boyle's *The Sceptical Chymist.* If Dippel had read this book, he would have learned that such a medicine could have included olives, bile, and even grapes.

Boyle wrote, "It seems then questionable enough, whether from Grapes variously order'd there may not be drawn more distinct substances by the help of Fire, then from most other mixt Bodies. For the Grapes themselves being dryed into Raysins and distill'd, will (besides Alcoli, Phlegm, and Earth) yield a considerable quantity of Empyreumatical Oyle, and a spirit of a very different nature from that of wine . . . The Juice of Grapes after fermentation will yield a *Spiritus Orders;* which if competently rectified will all burn away without leaving anything remaining."

What eventually became Dippel's Oil was used up until the end of the eighteenth century, when new and more powerful cures were found. Whether its users knew, or wished to know, what Dippel's Oil actually contained was a mystery, but this foul, odorous concoction was a mixture of ground-up animal blood and crushed bones, along with a few other ingredients—some human—that Dippel collected in a most unusual manner. Again, using human blood for curative concoctions was not unusual. Boyle insisted that "there is a Difference betwixt the saline spirit of Urine and that of Man's blood; that the former will not cure the Epilepsy, but the latter will."

And though Dippel's Oil did nothing for Dippel in the world of academia—the university appointments he had wanted did not materialize—his reputation as an alchemist grew. He even found favor in the royal courts.

This new interest in alchemy as a way to cure people also initiated in him a desire to study medicine. He chose as his place of learning the University of Leyden, in Holland.

With its long-standing tradition of printing and book trading, Leyden provided him with the ripe intellectual environment he had always craved. To this was added the presence of the city's university, the University of Leyden, the oldest university in the Netherlands, which was founded in 1575 by Prince William of Orange. As it happened, Leyden was also the place where the Leyden jar (a glass container insulated inside and outside with tinfoil capable of harnessing electricity) was invented in the mid-1700s, further reinforcing Dippel's link with the occult, electricity, and *Frankenstein*.

In Leiden Dippel came in contact with the great professor of medicine Herman Boerhaave. Though later Boerhaave became known for the disease that bears his name—Boerhaave syndrome, an illness that results in a rupture of the esophagus—at the time of Dippel's studies he was a celebrated professor of physics. Like Dippel, Boerhaave was the son of a clergyman who in turn had been eager to make a clergyman out of him. Like Dippel, he had also blended theology and medicine.

In the years that followed, Dippel came to believe that the gift of prophecy had been bestowed on him. As such, he set out to prophesize his own death, which he set for the year 1808. For someone who had been born in 1673, this was quite a stretch, giving rise to the rumor that perhaps he had rediscovered the philosopher's stone.

He was remarkably mistaken about that date, obviously demonstrating that not only was the philosopher's stone no longer in his possession, but that as a diviner his skills were lacking. On April 24, 1734, while enjoying the comforts of one of his patrons, the Count August von Wittgenstein, Dippel was found dead in his room, a peculiar bluish tint to his skin and foam collecting around his mouth. Immediately, those friends who saw the body believed that Dippel had been killed by one of his enemies. By then he had collected many, so one could only guess which enemy they were talking about.

The authorities were called, but those men, more than being suspicious, were superstitious; they had heard of Dippel's doings and the reputation he had gained. They were afraid to go near the body, let alone open it to perform an autopsy. They recalled that as they had made their way up to the castle, they had heard the rumor already circulating around the village that the devil had returned to claim Dippel's soul. Given that, the corpse remained untouched and the death attributed to "apoplexy."

IT REMAINS UNKNOWN IF PERCY SHELLEY AND MARY GODWIN heard of the infamous alchemist Johann Dippel, as Dippel is never mentioned in Percy Shelley's or Mary Godwin's diaries, nor is he spoken of in Mary's book *History of a Six Weeks' Tour Through a Part of France, Switzerland, Germany and Holland*. On returning to their boat, following their three-hour exertion on land, Mary continued writing in her journal in much the same vein she had done all along: "We were carried down by dangerously rapid currents, and saw on either side of us hills covered with vines and trees, craggy cliffs around by desolate towers, and wooded islands, where picturesque ruins peeped from behind the foliage . . . We heard the songs of the vintagers, and if surrounded by disgusting Germans, the sight was not as replete with enjoyment as I now fancy it to have been."

They were cruising on their way back to England, hopeful, yet unsure of how they would be welcomed.

Chapter 6

My Hideous Progeny

Whoever undertakes to set himself up as judge in the field of truth and knowledge is shipwrecked by the laughter of the Gods.

ALBERT EINSTEIN, *APHORISMS FOR LEO BAECK* (1953); REPRINTED IN *IDEAS AND OPINIONS*

O N April 10 and 11, 1815, Mount Tambora, the volcano on the Indonesian island of Sumbawa, began thundering loudly from within its crevice, unleashing one of the deadliest eruptions ever recorded in history. Fire rose from the cone-shaped crater, while lava flowed toward the surrounding villages. Pumice stones flew out of its depths, followed by a funnel of thick ash, which ominously rose up into the stratosphere and mingled with the water particles found therein.

This was not the first eruption the inhabitants had ever heard. Since nearly two years earlier—and most particularly in the last several months—the earth had often trembled beneath their feet and they heard a sort of inner-land growling, though they had ignored those signs; if anything, the inhabitants believed the volcano's belching was the gods who lived within it expressing their anger at the strangers who had intruded on their soil. Even years later, when the cause was attributed to natural phenomena, songs were sung about the vengeful gods wreaking havoc during the tragedy.

On the island of Batavia, the lieutenant governor of Java, Sir Thomas Stamford Raffles (who later founded the city of Singapore), heard the loud booms and thought they were gunshots. "The first explosions were heard on this island in the evening of 5 April, they were noticed in every quarter, and continued at intervals until the following days," he later wrote in his memoirs. "The noise was in the first instance almost universally attributed to a distant cannon; so much so that a detachment of troops were marched from Jocjacarta, in the expectation that a neighbouring post was attacked, and along the coast boats were in two instances dispatched in a quest of a supposed ship in distress."

The volcano had been dormant for nearly five thousand years, like many volcanoes around the world, and most especially those around the so-called Ring of Fire that made up the Indonesian archipelago, but great forces had converged beneath Tambora's surface and now they had built up to a magnificent and threatening level. By April 1815, the pressure had reached such a high point that when the final explosion occurred, the gigantic plume of gases and ashes traveled nearly twenty miles upward into the air. The larger chunks of debris that spewed out catapulted back onto the earth immediately, but many minute pieces remained up in the air, floating there for subsequent weeks and months.

The area surrounding the volcano suffered the most: "A violent whirlwind ensued which blew down nearly every house in the village of Sangir, carrying steps, or roofs, and light parts away with it. In the port of Sangir and . . . Tambora its effects were much more violent, tearing up the roots of the largest trees, and carrying them into the air, together with men, horses, cattle . . . the sea rose nearly twelve feet higher than it had ever . . . before . . . sweeping away horses . . . Every boat was forced from the anchorage," one of the few local survivors reported. The area was immediately plunged into pitch darkness, and the inhabitants, those thousands who had not perished in the explosion, were left without potable water, vegetation, or provisions of any sort.

When the volcano quieted, Lieutenant Raffles directed his men toward the Island of Sumbawa, where they were to gather information on the condition of the remaining villagers. One man who heeded the commands and reported back to Raffles was Lieutenant Owen Philips, who, after scouring the area, declared,

> *The extreme misery to which the inhabitants have been reduced is shocking to behold. There were still on the road side the remains of several corpses, and the marks of where many others had been interred. The villages almost entirely deserted and the houses fallen down, the surviving inhabitants having dispersed in search of food . . . Since the eruption, a violent diarrhoea has prevailed in Bima, Dimpo, and Sang'ir, which has carried off a great number of people. It is supposed by the natives to have been caused by drinking water, which has been impregnated with ashes; and horses have also died in great numbers, of a similar complaint.*

Though he did not know it yet, nearly 120,000 people had already lost their lives.

But Indonesia was not the only place that experienced the misery of Tambora. As the tiny ash particles traveled into the stratosphere, they were carried away by the west wind. Also the cloud of gas that had formed continued to rise to higher altitudes and merged with the vapor found there, forming sulfuric acid. As the wind blew this across the lands, it carried a faint sheen, a sort of thin mist that covered the lands. In later months, peculiar climatic events began to plague the entire planet, followed by the unusually rainy and cold summer of 1816, what became known as the Year Without a Summer.

As far away as the northernmost regions of Canada as well as in New England, frost persisted well into the summer, quickly destroying the entire season's crop for many farmers. In China the weather was blamed for destroying trees and rice fields and killing farm animals used for transport and provisions. The rains there caused the country's rivers and lakes to swell up, overflowing into villages and encampments and carrying with them cholera and diphtheria. Europe fared no better: Italy saw a peculiar yellow-tinted snow fall on its many villages and cities; potato fields died in Ireland, wheat in Germany, corn in France. Across the continent, prices rose with the riverbanks, as did riots, looting, anger, and violence.

But at the time no one knew precisely why these phenomena were occurring. Few had connected them to the Tambora eruption, because few had even heard about it. Whatever letters were sent from Indonesia to Europe or the Americas didn't arrive for weeks or even months, and by the time they did, Tambora was old news.

Aside from the many traumatic deaths, climate changes, and economic impacts, Tambora also triggered a strange effect on peoples' psyches. The long days of incessant rains, whipping winds, and shadowy and gloomy evenings, as well as the gray ashy snows, were not only alarming but also downright debilitating. For the members of Shelley's household, whose mental constitutions were already a bit weak, the foreboding weather only added to their woes.

Just as Tambora had experienced a series of inner blows culminating in a major catastrophe, Percy Shelley and Mary Godwin experienced a series of setbacks during the two years following their elopement. When they returned to England in September 1814, they discovered that their reputations had been tainted. Mary was ostracized by those who believed she had ruined a marriage, and by her own father, who was still as unyielding as an iron rod and refused to talk to them. She also learned at this time that she was pregnant. The pregnancy would have overjoyed Mary if she had not been preoccupied with the repercussions that came with it.

Mary had no official attachment to Percy. His marriage to Harriet Westbrook was still intact and a legal separation had yet to be discussed. She was only his mistress, so her child with Percy would be illegitimate and would not be afforded any of the legal benefits granted to a married couple's offspring. Both sets of parents also refused to assist them—financially as well as emotionally—as they set about building a family and a home together.

They also had to face the more practical and evident matter of Jane, or rather Claire, as she was now known. She had not only assisted but actually become a part of their scheme, and having had a taste of freedom, refused to return to the Skinner Street household, where her mother would keep a constant watch on her. The idea of working as a domestic did not appeal to her, nor did the prospect of joining a convent, as Mrs. Godwin had suggested; but she did not have the means to survive on her own, either. Only one solution remained, and that was to carve a place for herself in Mary and Percy's new home. This did not please Mary, and Claire knew it.

Yet, Claire made herself useful. When Mary's pregnancy made her unable to accompany Shelley on the various errands that took up most of his day—visits to "lawyers, insurance agents, and money lenders"—Claire took Mary's place during those ordeals. While this setup worked for a while, Mary soon became distrustful of the time Percy and her stepsister spent together, of the carriage rides they took, the long walks they enjoyed, of the teas they indulged in outside their home.

Claire had become a presence not only while Mary and Percy read, studied, or translated literary works, but also in their bed. Her temperament was easily influenced by outside words and tales, especially ones about ghosts, horrors, and devils, and she began to suffer nightmares after attending Percy's meetings with Hogg, as they often partook in such mystical conversations. Claire became uncomfortable listening to these stories and eventually developed a fear of being assaulted by ghosts resting on her pillows, phantasms traipsing to and fro about the house, and strange creatures cavorting in all manners of odious ways. Gasping for air beneath her blankets, seized by fear, she dashed out of her bed and into Mary and Percy's room, where she slunk between the two lovers and burrowed deep into the plush bed. Mary eventually believed that her stepsister's traumas were imaginary and only meant to break up her intimacy with Shelley. Shelley didn't mind at first but in time Claire's actions began to rub him the wrong way.

On February 22, 1815, Mary gave birth to a little girl, one that Dowden and others deemed a "seven months babe," though research suggests that the baby was born full term. The little girl began ailing almost immediately and was not expected to survive. For several days she gave Mary and Percy great delight, each small breath seeming perhaps like a small beacon of hope. But on March 6, Mary awoke to find her child dead. "Dream that my little baby came to life again," Mary later wrote in her journal, on Sunday, March 19. "That it had only been cold & that we rubbed it by the fire & it lived—I . . . awake & find no baby—I think about the little thing all day." That particular journal entry not only showed the depth of her sorrow, but also foreshadowed the direction of her literary endeavors—the idea of reanimation, in real life as well as in fiction.

Such scientific thoughts had played a major part during her

upbringing, but they had become even more magnified during her pregnancy. In the months preceding the birth, Mary, Percy, and Claire had often attended the Spring-Garden Rooms, where the brothers Garnerin had lectured on phantasmagoria, a show that was a blend of illusions, projectors, and parlor tricks commonly used in spiritual séances and that dealt with electricity and the properties attributed, or that could be attributed, to it. An advertisement that ran in *The Times* just over a month before the baby was born read, "Theater of Grand Philosophical Recreations [by Professor Garnerin] continues to be frequented by the most fashionable of the Nobility and Gentry. This week he will perform a great quantity of his finest experiments, entirely new, on the Gas, Electricity . . . Phantasmagoria &C."

These shows were not much of a novelty to the London community. Years before, at the dawn of the 1800s, Paul Philidor, who had displayed his innovations in Vienna and Berlin, had reinvented himself as Paul de Philipsthal and made a splash at the Lyceum Theatre in the Strand with his new version of phantasmagoria,

Grief-stricken over the loss of their child, Mary and Percy received even more heartbreaking news about Shelley's health. They consulted several doctors in London, and he was misdiagnosed with consumption and told that he was dying. No precise time had been given, but he was told he would not live more than a few months. All of this plunged Mary into an emotional turmoil. She craved comfort and consolation, but no one around her could provide that. Percy and Claire continued to spend a tremendous amount of time together, but did so outside of the home and Mary's sight. The Godwins did not offer their support because William Godwin still held a grudge, and Fanny Imlay was not allowed to visit as often as she wished.

Strangely enough, the only place Mary could find comfort was with Shelley's old friend Thomas Jefferson Hogg, to whom she sent a desperate letter: "My dearest Hogg—my baby is dead—will you come to me as soon as you can—I wish to see you—It was perfectly well when I went to bed—I awoke in the night to give it suck—it appeared to be sleeping so quietly that I could not wake it—it was dead but we did not find that out till morning—I am no longer a mother now."

Hogg hurried to visit and stayed until well into April.

MARY SHELLEY AND THOMAS JEFFERSON HOGG'S RELATIONSHIP HAS been dissected to bits since its infancy, though no one knows if it

Mary Shelley. This miniature portrait was made by Reginald Easton, and according to Lady Shelley, it was based on one of Mary Shelley's death masks. At her breast she wears a blue and yellow pansy, a symbol of remembrance she began to wear following the death of her husband, Percy Shelley, in 1822.

ever became sexual. The letters they exchanged during her pregnancy and following the birth of the baby show not only that Mary was a willing participant in their relationship, but also that she took pleasure from it. Betty T. Bennett, who edited *The Selected Letters of Mary Wollstonecraft Shelley*, even wrote, "The degree of her involvement remains a matter of speculation, though those letters strongly suggest that the affair between Mary Shelley and Hogg was not consummated."

If an affair did take place, Percy Shelley knew about it and even blessed it. He had insisted on Mary and Hogg's intimacy, perhaps because during this period—during Mary's pregnancy of 1814–15—he and Claire Clairmont began their own sexual affair. If so, it explains Mary's antipathy toward her stepsister.

In any case, Mary certainly knew that Hogg was in love with her, or at least believed himself to be, and her pregnancy allowed her to stall his advances. In a letter she wrote to him in January 1815, she said, "You love me you say—I wish I could return it with the passion you deserve—but you are very good to me and tell me that you are quite happy with the affection which from the bottom of my heart I feel for you . . . But you know Hogg that we have known each other for so short a time and I did not think about love—so that I think that *that* will come in time & then we shall be happier."

Years later, an elderly Claire revealed to an interviewer named Edward Augustus Silsbee that Shelley actually wanted Mary to have sex with Hogg. In a memorandum book, following a conversation with Claire, Silsbee jotted down: "C's story of Mrs. S's coming into her room about when they lived was S. Pancras Arabetta—& putting her head on her pillow . . . Saying Shelley wanted her to sleep with Hogg."

Caught up in this peculiar love scheme, the death of her daughter, a lover doctors believed was about to die, a father who did not want to see her, and constant money worries, Mary could see that the happiness she had sought with Shelley while conversing on the grounds of St. Pancras was eluding her and her life was turning into a bleak existence reminiscent of her mother's. And as she sank deeper into her melancholy, Claire also realized that she had reached a juncture in her own life. She knew she could not live at Skinner Street or in Mary's home, where the hostility had now became palpable. She had also decided it was time to find a lover of her own. And she wanted to snag someone who was more renowned than Percy Shelley, and who would eventually eclipse the poet's notoriety. Even though Mary was praised for her intellect and Fanny respected for her piety, Claire knew that in her case it was her physical beauty people admired. The mass of her dark ringlets, the olive tone of her skin, the roundness and softness of her curves, and her sociable disposition, she believed, would be enough to catch a man's attention. With that in mind, Claire set out to seduce the poet George Gordon Noel Byron, better known to the world as Lord Byron.

BYRON ALREADY HAD A REPUTATION AS A LASCIVIOUS, SHOCKING madman. He was the son of Captain John Byron, a man who later became known as Mad Jack, given his enjoyment of women, drink, and poker. And people agreed that family tradition had continued on, particularly in regard to women. Although many individuals felt his antics were repugnant, his beauty and charm were such that women of all ages and social status coveted him, and even men, the ones who should have loathed him most, ended up admiring him.

Thomas Moore once said about him:

Of his face, the beauty may be pronounced to have been of the highest order, as combining at once regularity of features with the most varied and interesting expression . . . His eyes, though of a light gray, were capable of all extremes of expression, from the most joyous hilarity to the deepest sadness, from the very sunshine of benevolence to the most concentrated scorn of rage. But it was in the mouth and chin that the great beauty as well as expression of his countenance lay . . . The glossy dark-brown curls, clustering over his head, gave the finish to its beauty. When this is added to his nose . . . that his teeth were white and regular, and his complexion colorless, as good an idea as perhaps it is in the power of mere words to convey may be conceived of his features.

One of Byron's most notorious lovers had been Lady Caroline, the woman who came to think of him as nothing more than "mad, bad, and dangerous to know." Already married to a man who would later become part of London's Parliament, Lady Caroline was so enraptured and infatuated with Byron she had hatched a plan to run off with him. But Byron scoffed and declined the invitation.

Byron was never amused by women's antics or their girly talk, and even less by their suggestive smiles and batting eyelashes. He despised the clinginess and neediness of most of his lovers and was even less drawn to their serious desire to attach themselves to him, which he could not understand, because even he knew he could not be counted on. "You will find that I am the most selfish person in the world," he once told an acquaintance. "I have however, the merit, if it be one, of not only being perfectly conscious of my faults, but of never denying them; and this is surely something."

He especially enjoyed his reputation for having illicit and even scandalous affairs. This love of the forbidden even drove him toward his half sister Augusta Leigh. Five years his senior and also the daughter of Mad Jack, she too had inherited some of the old man's ability to shock people. Indecency was also one of Augusta's most remarkable traits. The daughter she eventually gave birth to, Medora, was said to have been fathered by Byron. Instead of being scandalized by this, Byron believed such a disreputable idea only added to his coat of infamy.

Many believed Byron would never marry because having a wife would thwart his unnatural desires. But thanks to Lady Melbourne, Lady Caroline's mother-in-law, he was introduced to Annabella Milbanke, who eventually became his wife. Any notion that Annabella possessed the strength to tame Byron was laughable to most. Still, they were married, even though the union was doomed from the start.

Byron possessed several traits that appeared not merely mad but downright psychotic. They were due not to mental deficits, but to his incessant drinking, coupled with his ferocious anorexia. He was a compulsive dieter, often existing on a regimen that consisted of a few cookies and copious amounts of green tea, which he usually consumed in early to midafternoon. Occasionally he added a few vegetables. He also used a massive amount of purgatives to maintain what he believed was his ideal weight. As such, he was always "extremely thin, so much that his figure [had] almost a boyish air," further enhanced by the clothes he wore, which hung from his body.

This fear of gaining weight was tied to his desire to look thin, but also to what he believed was a mental superiority he got from his ability to control what he ate. He once told Lady Blessington that if he gave in and ate more food, "he should get fat and stupid, and that it was only by abstinence that he felt he had the power of exercising his mind."

He felt that if he got carried away as others did, all of his "intellectual faculties would dwindle." He said the regimen "made him feel lighter and more lively." On one occasion he told Moore that "the devil arrives with plumpness, and I must drive him away through hunger! I DO NOT WISH TO BE A SLAVE OF MY APPETITE." But his alcohol consumption, poor diet, and views on weight gain caused his worst traits to come out.

He was also highly self-conscious of a limp he developed in his youth due to a congenital defect. He obsessed over the limp and concealed it so successfully that few people ever noticed, though doing so often brought him great pain after long walks. He never forgot being taunted by his schoolmates because of the "lameness." This, he mused, was "the greatest misfortune, one that [he was] never able to conquer." He knew that a person needed a certain mental aptitude to "conquer the corroding bitterness" that arose from such a situation, and that he did not have that aptitude.

He had devised a way of walking on his toes that not only lengthened his stature (he was of average height—five feet eight inches tall), but also allowed his limp to be diminished or disappear. That he spent unusual amounts of time practicing and developing those abilities seemed repugnant to many outside of his immediate circle. His wife was one of those people who over time came to believe he was insane. But he was not medically insane; many people agreed that she lacked the sense of humor he most prized and that she needed to understand him.

Byron had also developed a habit of telling tall tales, with him-

self usually as the principal character in them. Those tales were
meant to impress and most people realized they were embellished.
Lady Byron was too late in figuring out when he was joking and
when he was not: "He had wished to think [himself] partially de-
ranged, or on the brink of it, to perplex observers and prevent them
from tracing effects . . . ," she wrote in a letter republished in *The
Passages from Lady Anne Barnard's Private Family Memoirs.* "By the introduction of fictitious incidents, by change of scene or time, he has
enveloped his poetical disclosures in a system impenetrable except
to a very few." Despite the discrepancies in their characters, in December 1815 Lady Byron gave birth to a baby girl. They named her
Augusta Ada, after Byron's sister, though Annabella's family always
referred to her as Ada. Barely a month later, Annabella left Byron.
Following the separation, and for years afterward, he said he could
not understand why she left him. Many felt this was an act.

Though neither spoke of the real reasons for their separation, it
was speculated that Byron had told his wife about his real involvement with his half sister, and she couldn't forgive him. He wrote
to her often, but those letters always went unanswered. For nearly
a year after the separation, Byron admitted to her, in a letter he did
not send, that he hoped for a reconciliation.

Although his marriage to Annabella had really been a sham,
Byron had felt affection for his new daughter. "The little girl was
born on the 10th of December last," he wrote in a letter. "Her
name is Augusta Ada (the second a very antique family name, I
believe not used since the reign of John). She was, and is, very
flourishing and fat, and reckoned very large for her days. Squawks
and sucks incessantly . . . her mother is doing very well, and up
again. I have now been married a year on the second of this month,
heigh-ho!"

The loss of not seeing the baby every day now disappointed him, as he knew he would not be a part of her growing up. He despised the notion that Lady Byron would get to "feast on the smiles of her infancy and growth," though he vowed that "the years of her maturity" would be his. Soon after the breakup of their marriage, rumors began to swirl about of Byron's violence, sodomy, and illicit affair with his half sister. He could not understand why people were so interested in the separation of a husband and wife, something that was common, he thought. It was during this time that he began receiving letters from Claire.

Initially, she wished only to meet him to discuss opportunities for her future. She had been on the lookout for ways of becoming independent and believed she could do so as a stage actress. Byron was a member of the Drury Lane Theatre, and Claire believed he could assist her: "May I beg you . . . if it is not too difficult to procure from one of your theatrical friends an account of what instructions are necessary for one who intends entering that career. What are the first steps to be taken . . . Is it absolutely necessary to go through the intolerable & disgusting drudgery of provincial theatres."

The letters soon turned sexual, especially when he began calling her a little "fiend." In March or April 1816, she wrote him the first letter in which she divulged her unattached state as well as some of her deeper feelings: "If a woman whose reputation had yet remained unstained, if without either guardian or husband to control she should throw herself upon your mercy, if with a beating heart she should confess the love she has borne you many years, if she should secure to you secrecy & safety, if she should return your kindness with fond affection & unbounded devotion, could you betray her, or would you be silent as the grave?"

Byron must have seen Claire as a diversion from his daily routine, though Claire must have recognized it as a good way to upstage her sister. Though he made her no long-lasting promises, he could introduce her to a large repertoire of sexual experiences, and Claire dived in with abandon, as he exerted a strong sexual allure over her. Even years later, when nearing her death, she thought of the time she had spent with him as nearing the sublime. But while Claire enjoyed her newfound sexual freedom and all of Byron's physical favors, quickly falling in love with him, Byron thought of her as nothing more than an enjoyable interlude. Soon enough, the relationship came to its inevitable denouement.

Byron had decided to leave London for the Continent, wanting to salvage what was left of his reputation and what was left of his physical and mental health. He had planned on Switzerland, traveling across Europe, down the Rhine River, until he reached the magnificent shores of Lake Geneva.

Claire heard about his plans and decided to follow him, though she had assured him she would not do so: "I assure you nothing shall tempt me to come to Geneva by myself since you disapprove of it as I cannot but feel that such conduct would be highly indelicate." But in the same letter, she almost warned him that should she find someone willing to accompany her, she would go, as she was certain it "would not displease [Byron]." In other words, she planned on meeting him in Geneva not only because she was deeply in love with him, but because she had also just learned she was carrying his child. Believing that he not only loved her but that he would also take care of her and their child, Claire realized she needed her sister's assistance.

On January 24, 1816, Mary had given birth to a son, William, named to honor her father. Though this new birth brought joy, everyday events still dominated their lives: There was Claire, the now-constant companion in Mary and Percy's lives. And money, or the lack of it, continued to plague them. As a consequence, money-lenders were also continuously after them, and fearing that Shelley would be arrested for their debts, they decided to leave England for Italy. They believed the trip would take them away from the clutches of the loan sharks and would remove Shelley, with his fragile health, from London's dreariness. And it was supposed to give Mary a respite from her concerns. Claire, of course, wanted to go to Switzerland, not Italy, and made her opinions known. Following several conversations, they all agreed on Lake Geneva, a sojourn, they hoped, that would bring about much-needed tranquility and mark a new era in their lives.

While Mary, Percy, their child, and Claire prepared for their journey, Byron also made arrangements for his own separate departure. He commissioned a carriage to be built, something as pompous as the one belonging to Napoleon, possessing "a library, a plate-chest, and every apparatus for dining." He made certain his friends knew where to find him, and he also hired a physician to ride along for the journey, someone who would look after his physical and mental health as he sometimes seemed incapable of doing himself. He gave this impossible and irrational task to the young, naïve, and vain Dr. John William Polidori.

DR. JOHN WILLIAM POLIDORI WAS A YOUNG MAN OF TWENTY WHEN he was recommended to be Lord Byron's physician. The job could not have come to him at a better time: though a medical doctor in his own right, having attended the prestigious University of Edinburgh Medical School (where he had earned his degree as one of the youngest physicians ever), he had never truly wanted to be a doctor but had become one at the urging of his father, Gaetano Polidori, a domineering man who ruled with a strict hand. The young Polidori had leaned instead to the religious life, as well as the literary one.

This trip with Byron seemed like not only an opportunity too good to pass up but also one that could give him a new calling. The job involved traveling beyond the confines of England and Scotland, but also recognition and attention, and a chance to try his hand at a literary career. He believed, somehow, that by being in Byron's company that could be easily achieved.

Polidori's father was against his son's traveling with Byron. The idea of his son's fraternizing with such an individual struck a note of terror in Gaetano, and he had tried to dissuade Polidori, but he got nowhere. He had heard of Byron's reputation and feared his son would be badly influenced by it. Despite Gaetano's voicing his misgivings, young Polidori refused to listen.

Polidori already came from a literary family, members of which had made, and would continue to make, their marks in the field. His father, originally a lawyer, had been a secretary to the famous and infamous writer Vittorio Alfieri; he had also taught Italian in England and translated many literary and critical works. And one of Polidori's sisters, Frances, would go on to marry another Italian, Gabriele Rossetti. The couple would give birth to a brood of children, one of whom would become known in literary circles as the poetess Christina Rossetti, and another who would make a name as a poet and a painter, Dante Gabriel Rossetti.

When he was recruited for the trip abroad, Polidori also received another cushy assignment: he had been hired by Byron's publisher, John Murray, to keep a diary, a journal of his travels with Byron. In it, Polidori was to catalog details of what they saw and visited as they crossed the various countries, of the changing landscapes and picturesque rides down rivers and valleys, of the people they encountered, of the details of taking care of Byron's health, as well as the changing moods of the poet's capricious nature. Titled "Journal of a Journey Through Flanders," Polidori's travelogue was to be published upon his return to London. Few knew he was keeping track of his—and Byron's—doings, not even Percy Shelley and his new family, who would become subjects of the journal as well.

The journal has had a very convoluted history. It spanned from April 24, 1816, through December 1816, when Polidori abandoned it. It was, and is, an interesting read because Polidori's diary has remained one of the only written records, albeit in short passages, of the famous ghost story competition that gave birth to *Frankenstein,* or so Mary Shelley said it did. Polidori began his journaling on the day of their departure from London, keeping if not meticulous notes, at least a somewhat scanty log of his doings and his impressions of Percy Shelley, Mary Godwin, and Claire Clairmont, which were not always positive.

But Polidori's words have to be carefully scrutinized. Upon his death the diary ended up in his sister's hands. Charlotte Lydia Polidori then set it aside for later scrutiny and did not read it until she was an older woman. What she read scandalized her. Many passages—those that spoke of Byron's excessive sexual appetite and of her brother's jaunts to a bordello (which usually ended with a visit from the local police)—she found too "salacious." They were improper, and she would not have her brother's name associated with them. So she transcribed the diary onto fresh pages, omitting the more raunchy passages. She then handed this new edition over to her nephew William Michael Rossetti and burned the original diary. Rossetti, who eventually edited Polidori's diary, assured readers that "the authority is only a shade less safe than that of the original," though in reality he could not have been sure of Charlotte's accuracy.

PROBLEMS BETWEEN BYRON AND POLIDORI AROSE RIGHT AWAY because Polidori believed he was traveling with Byron as a companion, a friend who was on par with Byron, and not as a doctor. In a letter Polidori sent to his sister, this became painfully obvious when he declared proudly, "I am very pleased with Lord Byron. I am with him on the footing of an equal, everything alike . . . He has not shown any passion; though we have had nothing but a series of missteps that have put me out of temper though they have not ruffled his."

The cold, misty, and ashy-gray weather as they crossed the channel often caused temper tantrums in the group, and Polidori's only description of the other passengers on board was to say they "looked dreary." He had yet to experience the excitement he had envisioned for the journey.

For Byron, the changes of scenery appeared to be doing wonders already, especially in regard to his sexual nature. "As soon as he reached his room," Polidori commented, "Lord Byron fell like a thunderbolt upon the chambermaid."

When they reached Switzerland, Lord Byron learned that a note had been left for him at the local post office. It was from

Claire, who was writing to tell him that she, along with Mary and Shelley, had already arrived and were staying in the area. She also seemed disturbed that he was not already there and feared he may have lied about coming to Geneva in the hopes of avoiding her. The impassioned letter, Polidori wrote, was "worthy of a novel." Lord Byron had not planned on seeing Claire while in Switzerland, and he wasn't pleased that she was there. Though he had initially been charmed by her features, which were bewitching in their exoticism, she was now encumbering him. He was always pestered by English expatriates when he was traveling abroad. It did not matter to him that Claire had been his lover.

The Shelley party had not departed London until May 3, eventually arriving in the Sécheron neighborhood of Geneva nearly two weeks later. They had retraced part of the route they had taken two years earlier, but this time they had a child with them. And as before, Mary was struck by the dullness of the French.

Mary's and Claire's lives now revolved around lovers, affairs, children, all of which distracted them from worrying about their sister Fanny Imlay. Quiet and prudish, she had now reached the age of twenty in the dingy Skinner Street household, under the naughty eye and wagging tongue of Mrs. Godwin. Fanny had no romantic entanglement of her own, nor any real prospects; she did not inspire love, lust, or maddening outbursts or passions, and she knew it. Not an intellectual, she received no stimulation from books or translations, nor was she taken to see the electrical shows and lectures on phantasmagoria.

She existed in a sort of limbo, her life standing as still as a murky pond. Far removed from her sisters' lives on Lake Geneva, she read and reread the letters she received and lived voyeuristically through their adventures, travels, and doings. She often replied to Mary's letters and asked about the people they were with, most especially Lord Byron, whose reputation she, like all of England, had heard of: "Does he come into your house in a careless friendly dropping in manner," she asked in a July letter. "I wish to know though not from idle curiosity whether he was capable of acting in the manner that the London scandal-mongers say he did?" Still on the topic of Byron, she said: "I cannot think that from his writings that he can be such a detestable being—Do answer me these questions! For were I to love the poet I should like to respect the man."

There had been talk of Fanny's moving to Ireland with her aunts Everina Wollstonecraft and Eliza Bishop—her mother's sisters—but those two women did not seem willing to take her in. Various reasons were put forth, but the most probable was that they despised William Godwin and did not want to make his life easier, which taking Fanny would have done. But Fanny was the one to suffer. Not that Ireland would have been the perfect situation either: in Dublin, Eliza and Everina ran a school with a strict code of conduct, Eliza teaching the girls and Everina the boys. Fanny's life there would have been just as glum as it was already at Skinner Street.

Fanny occupied a tricky spot in the household—part mediator and part referee between her sisters and the rest of the family—and the attempts to find balance were trying and tiresome. This also gave Mrs. Godwin a chance to show her petty and vengeful side by often gleefully telling Fanny that her sisters were making fun of her. Unfortunately, Fanny came to believe Mrs. Godwin's view that her life was going nowhere. In time, Fanny came to think that indeed she had become an object of their mockery, though in reality nothing of that sort had occurred. It was only a matter of time before Mary Wollstonecraft's melancholy and depression began to show themselves in Fanny.

But Mary, Claire, and Percy did not notice anything wrong with Fanny. Instead, Mary wrote her lengthy letters about the European landscape, and upon arriving at Poligny she included accounts of her pleasant and not-so-pleasant encounters; of cities ablaze with life and grandeur; of lofty and obscure forests populated by the tallest trees ever seen that seemed to guard their own secrets; of a lake so blue that like the tantrums of a child changed its colors at any moment, from cerulean blue, to hazel, to gray, depending on the time of day. "The town is built at the foot of the Jura," she wrote, "which rises abruptly, from a plain of vast extent. The rocks of the mountains overhang the houses."

One reason they left England was to find more suitable weather for Shelley, and to their unhappiness, spring had not yet arrived on the lake. The villagers told them the season had dawned unusually cold and rainy, which they soon discovered themselves. "As we ascended the mountains," Mary wrote to Fanny, "the same clouds which rained on us in the vallies poured forth large flakes of snow thick and fast." The desolation and coldness of the weather seemed to permeate the village and did nothing to aid Mary's temperament. "Never was a scene more awfully desolate. The trees in these regions are incredibly large, and stand in scattered clumps over the white wilderness; the vast expanse of snow was chequered only by those gigantic pines."

They scoured the lake region for a place to stay and finally settled on Maison Chapuis, a short walk from Villa Diodati. There,

Byron had set up residence in a sprawling estate that was too large for his personal party but large enough to contain his reputation.

The weather followed a certain pattern: it moderated during the daylight hours, while the evenings were punctuated by thunderstorms and lightning. The group took advantage of the morning hours and spent them on the lake. The air was perfumed with the scents of newly bloomed flowers and the still-frigid waters matched the crisp air. Mary soon began to revel in her surroundings and a change occurred in her. "You know that we have just escaped from the gloom of winter and London," she wrote again to her sister, "and coming to this delightful spot during this divine weather, I feel as happy as a new-fledged bird, and hardly care what twig I fly to, so that I may try my new-found wings. A more experienced bird may be more difficult in its choice of a bower; but in my present temper of mind, the budding flowers, the fresh grass of spring, and the happy creatures about me that live and enjoy the pleasures are quite enough to afford me exquisite delight."

As the English expatriates settled on the banks of the lake, a peculiar dichotomy soon arose among them. Byron was drawn to Shelley, and the two poets enjoyed spending time together, touring the lake and the region around it, the scattering of villages on its shores, bantering with each other and discussing philosophical subjects. More often than not, Byron expressed his less-than-praising opinions of his fellow Englishmen. He liked stories that let him shock those with a more delicate disposition. Anyone who knew him knew he liked to be scandalous, but others could be surprised by his candor, particularly when he talked about his own love life and the women he had bedded.

According to Claire Clairmont, who spoke of those months on the lake later in life, Byron often talked of his half sister Augusta, detailing the relationship they had enjoyed and that she had given him two children. Though Claire and Mary were stunned, Shelley had already had the experience of reassuring the women that Byron was out to startle them and nothing more. A good example of this was Bryon's tale of a lover in Constantinople whom he had had killed. As Byron recalled, the young woman had been unfaithful to him and as an act of revenge, he had hired a killer to dispose of her; this hired man had stitched the woman inside a fabric sack, then thrown her into the water and allowed her to drown.

The rest of the group seemed to hover around the two poets. Whenever Polidori tried to inject himself between the two of them—always too eager to have his literary works discussed and to get in Byron's good graces—he was quickly rebuffed. He became increasingly jealous of Shelley, who he believed had stolen away Byron's attention. With time on his hands, Polidori often frequented the bordellos and gambling halls that bordered the lake.

Mary, on the other hand, felt abandoned by Shelley and began to spend her time with Polidori, whom she saw as a younger brother, even though he was older. And Claire found that whenever she tried to see Byron, she was derailed by either Shelley or Polidori, one of whom was always in the way. Claire often snuck away from Maison Chapuis at night to meet Byron but came face-to-face with a vigilant Polidori. She eventually came to despise the doctor and had no qualms about expressing those feelings. She tried to convince Byron to dismiss the young doctor, but Byron seemed to take a certain pleasure in her despair. "Pray if you can send M. Polidori either to write another dictionary or to the lady he loves," Claire wrote Byron in a moment of heat. "I hope this last might be his pillow & then he will go to sleep; for I cannot come at this hour of the night & be seen by him; it is so extremely suspicious."

The villas and cottages surrounding the lake were occupied by local residents, but also by English expatriates who had heard about Byron's sexual perversions as well as Shelley's elopement. Often, while clamoring for tidbits, they directed their expensive telescopes toward the group's villas, hoping to catch a glimpse of something disreputable going on. On one occasion colorful table-cloths that had been left to dry in the open air were mistaken for the girls' petticoats. These eavesdroppers buzzed with gossip about the strange and scandalous ordeals taking place. They spun saucy tales to while away the sleepy afternoons, and some of them even reached all the way back to London.

Byron, Shelley, Mary, and Claire were well aware of the rumors being spread, and not surprisingly Byron enjoyed them. Some years later, in a letter Percy Shelley wrote to Countess Guiccioli, Byron's Venetian lover, he made his true feelings known:

> Our dwellings were close together; our mode of life was quiet and re-tired; it could be impossible to imagine an existence simpler than ours, less calculated to draw down the aspersions cast upon us . . . Both Genevans and English established at Geneva affirmed that we were leading a life of the most unblushing profligacy. They said that we had made a compact together for outraging all held most sacred in human society . . . I will only say that incest, atheism, and many other things equally ridiculous or horrible were imputed on us. The English news-papers were not slow in propagating the scandal, and the notion lent entire faith. Hardly any mode of annoying us was neglected. Persons

*living on the borders of the lake opposite Lord Byron's house made use
of telescopes to spy out all his movements. An English lady fainted, or
pretended to faint, with horror on seeing him enter a saloon. The most
outrageous caricatures of him and his friends were circulated; and all
this took place in the short period of three months.*

During the warmer parts of the days that summer, they
boarded sailboats on the lake and tracked the various paths that
lined the area, or they took excursions to the neighboring villages.
But as the afternoons grew dark and the evenings even darker, the
temperatures dipped, the thunderstorms plaguing the Continent
due to the volcanic eruption still made themselves heard beyond
the mountain ranges, and the yellow streaks of lightning sliced
through the inky-black night.

That was when they convened at Villa Diodati around a blaz-
ing fire, its pleasing warmth inflaming their own imaginations.
The whining of the wind made its way through the forest and the
stormy weather pummeled the windowpanes, perpetuating a sense
of mystery that directed the topics of their conversations.

"The thunders that visit us are grander and more terrific than
I have ever seen before," Mary wrote to Fanny. "We watch them
as they approach from the opposite side of the lake, observing the
lightning play among the clouds in various parts of the heavens,
and dart in jagged figures upon the piney heights of Jura . . . One
night we enjoyed a finer storm than I had ever before beheld. The
lake was lit up—the pines of Jura made visible, and all the same
illuminated for an instant, when a pitchy blackness succeeded, and
the thunder came in frightful over our heads amid the darkness."

Thunder and lightning were already a part of their lives. Mary had been born during a thunderstorm, and Shelley was already familiar with the attributes of such atmospheric phenomena. Sitting in the shadow of Lake Geneva, they were all aware of Benjamin Franklin's, Humphry Davy's, and Luigi Galvani's works, the scientists' notions of the spark of life. Viewing and hearing the natural spectacle outside their windows, they could not help but converse about such possibilities, about the notion of reanimating the dead.

IN THE 1831 EDITION OF *FRANKENSTEIN*, MARY SHELLEY WROTE IN HER introduction: "Many and long were the conversations between Lord Byron and Shelley, to which I was a devout but nearly silent listener. During one of those, various philosophical doctrines were discussed, and among others the nature of the principle of life, and whether there was any probability of its ever being discovered and communicated . . . Perhaps a corpse could be re-animated; galvanism had given token to such things; perhaps the component parts of a creature might be manufactured, brought together, and imbued with vital warmth."

These words made it appear as if the conversations took place over an extended period of time and only between the two poets. That meant the rest of the group just stood by and listened. Polidori's diary recounted something different.

In an entry dated June 15, Polidori noted, among other details, "Shelley, etc. came in the evening; talked of my play etc., which all agreed was worth nothing. Afterwards Shelley and I had a conversation about principles—whether man was thought merely an instrument."

After so many years, is it possible that Mary had simply forgotten about the other two people, Polidori and Claire, involved in the conversation on galvanism and the restoration of the dead, which in turn gave rise to her story? Whether or not the discrepancy occurred unintentionally, it was more likely that Dr. John William Polidori had been the one debating with Percy Shelley, not Lord Byron.

Shelley's knowledge about science, the occult, and medicine had been gained by reading various texts famously published at the time, but also from personal experience. Following his expulsion from Oxford University, he had spent some months with his cousin Charles Grove attending Joseph Abernathey's anatomical lectures in London. He had even debated becoming a surgeon. Polidori's experience with these matters was real and tangible. He had graduated at the top of his class from one of the most distinguished medical schools in Europe: the University of Edinburgh Medical School, where his classes had included "Anatomy, Surgery, The Theory and Practice of Medicine, Chemistry, Botany, Pharmacy, and the related discipline of Materia Medica."

The institution was famous for churning out well-known physicians, but it also had a long-standing tradition of body snatching, which Polidori might very likely have recounted to the party, given that the subject fit with their evening conversations. It was also well known that a student who wished to attend that medical school but did not have the financial means could cover his tuition by providing professors with dead corpses to dissect. Thus, all over Edinburgh the graveyards were looted; doctors and lecturers always knew where the corpses were coming from, and the citizens also knew who was robbing the graves. It's not known if Polidori was a grave robber himself, but he had certainly heard of such behavior because its notoriety spanned the country and beyond its borders.

It is also all but certain that he worked on and examined bodies robbed from graveyards. Of particular interest to Polidori during his studies was somnambulism, or motor action, such as walking, during sleep. He was most especially interested in the philosophical attributes tied to the condition and how that related to the so-called principle of life. His studies on the subject did not make him an expert, but he had enough knowledge to write about it and discuss it with some authority.

In 1815 Polidori had finished his medical studies with a thesis titled "Oneirodynia," a Greek word meaning "waking while in a dream," or in better terms, waking-dream, the very state Percy Shelley often suffered from and to which Mary Shelley always attributed the arrival of her story *Frankenstein; or, The Modern Prometheus.* Polidori's thesis was not exceptionally long or scientific, and most of the writing was based on case histories from his uncle Aloysius Polidori, a doctor himself. Its pages make it clear that Polidori's interest was in the separation of the mind and the body, which many believed could occur during sleep.

He went to great lengths to describe the case of a ten-year-old boy who had been taken down by frequent headaches, and "whose paternal aunt was prone to epileptic insults." It seemed that the boy was also prone to "chronic convulsions, a tremor of the knees, which was followed by the collapse of his body to the ground, pain in the head, and finally sleep."

The boy's symptoms worsened and were only relieved when bloodletting was applied to him. Soon, though, he was also "chattering and gesticulating," and had a vision of "French men . . . trying to attack him." But of particular interest to Polidori's theories was that while those attacks were being experienced, if a flame was placed or swayed before the boy's eyes, which were open, he did not flinch or push it away, clearly indicating that his physical body was in one place, while his mind was somewhere else. To make matters worse, the boy's thirteen-year-old sister and a friend of hers of the same age soon began to suffer from the same symptoms.

While Polidori used the boy's case and others similar to his to validate and illustrate his research, he had neglected to consider the possibility that the subjects might have been lying, or simply embellishing what happened to them.

The conversations at Villa Diodati would not only have been on par with Polidori's level of understanding and knowledge, but they also would have offered him a chance to impress the group, most especially Mary Godwin, on whom he had developed a sort of amorous crush.

JUNE 15 HAD NOT BEEN A GOOD DAY FOR POLIDORI. EARLIER THAT morning, while perched on a balcony overlooking the lake, he had been urged by Lord Byron to hoist himself up and jump off its parapet. The Shelley party had been heading toward Villa Diodati, and Byron, who had become aware of the younger man's feelings for Mary and of his excessive sensitivity, had assured Polidori that such an acrobatic act would impress her. It did not, and he ended up spraining his ankle, which became more painful when Byron persuaded him to read aloud one of his plays. His rendition of the play, and the play itself, were thought of as useless and were relentlessly ridiculed for hours.

This must have made him feel like the resounding failure his father had warned him he would become. Even worse, Polidori had

also become the butt of their jokes, most especially at the hands of
Byron, who had nicknamed him "Polly-dolly." Having been dis-
paraged and mocked for his writing as well as for his painful and
flirtatious jump, it made sense that, although wounded, he would
have inserted himself in a conversation about the principle of life
or somnambulism, because at least he knew more about the subject
than the rest of them.

On that particular evening—June 15—and over the next three
days, the famous ghost story competition to which *Frankenstein* has
always been, in part, attributed took place. On June 17, Polidori
noted in his diary, "Went into town; Dined with Shelley, etc . . .
the ghost-stories are begun by all but me."

Everyone else, including Mary, was already hard at work on
their tale. That particular detail, however, contradicts slightly the
account Mary later gave in her introduction to *Frankenstein,* where,
in unflattering terms, she wrote, "Poor Polidori had some terrible
idea about a skull-headed lady who was so punished for peeping
through a hole . . . I busied myself to think of a story . . . one which
would speak to the mysterious fears of our nature and awake
thrilling horror—one to make the readers dread to look round, to
curdle the blood, and quicken the beating of the heart . . . 'Have
you thought of a story?' I was asked each morning, and each morning I was forced to reply with a mortifying negative."

In Mary's version of events, she was the last to begin creating a tale, and she gave readers of her introduction the impression that the unfolding of her story took place over a substantial
length of time, rather than a handful of nights. She also suggested that the participants in the contest shared their stories
with each other as they were set down to paper, and that they

were discussed and dissected around the fireplace. But that was not true. No one spoke of his or her doings; no one knew what the others were writing about. Several years later Mary Shelley herself agreed that the details of their stories were always kept most secret. But if that was so, how did she know what Polidori was writing about? Not only that, Polidori had never mentioned a skull-headed lady. He always maintained that the fragment he came up with during those evenings gave birth to his story "Ernestus Berchtold."

The group's interest in writing their own ghost tales had not only sprung from their conversations about galvanism and the possibility of reawakening the dead, but also from the reading they were indulging in. They were all fascinated by *Das Gespensterbuch*, the voluminous collection of German ghost tales translated into French by Jean-Baptiste Benoît Eyriès. Retitled *Fantasmagoria*, the text was one of a handful of books Lord Byron had requested from his publisher in London. Populated by vampires, spirits, and tales of unbridled passion and unfulfilled love affairs, the stories inspired them to try their own hands at writing similar stories and must have prompted Byron to whisper the now-infamous words, "We will each write our own ghost story."

The reading continued on the next evening. Byron must have leafed through a collection of poetry he had on hand, including a volume of Samuel Taylor Coleridge's that had been published some weeks before by John Murray; this included the rhyming poem *Christabel*. As Mary, Percy, Claire, and Polidori gathered round a fire, listening to the wind whip across the lake, Byron must have removed the volume from his pile, and speaking in that tone Lady Blessington remembered as "neither low or high," began:

The lady leaps up suddenly,
The lovely lady, Christabel!
It moan'd as near, as near can be,
But what it is, she cannot tell—
On the other side it seems to be,
Of the huge, broad-breasted, old oak tree.

The howling thunder outside the windows seemed appropriate accompaniment for the tales they were telling:

Sweet Christabel her feet she bares,
And they are creeping up the stairs;
Now in the glimmer, and now in the gloom,
And now they pass the Baron's room,
As still as death with stifled breath!
And now have reach'd her chamber door;
And now with eager feet press down
The rusties of her chamber floor.

As lightning from the storm got nearer, it boomed across the lake and heightened the group's mood—most of all that of Shelley, who seemed entranced by the cadence of the lyrics, by the piercing thunder that weaved relentlessly among Byron's words, awakening a curious sensation in the listeners:

Beneath the lamp the lady bow'd
And slowly roll'd her eyes around;
Then drawing in her breath aloud,
Like one that shudder'd, she unbound

The cincture from beneath her breast;
Her silken robe, and inner vest,
Dropt to her feet, and full in view,
Behold! Her bosom and half her side—
A sight to dream of not to tell!
And she is to sleep by Christabel.

Shelley fractured the silence by leaping from his chair and running out of the room. It was unclear what had frightened him—the poem, the mood in the room, or the weather outside the windows—but in the frenzy that followed, Polidori rose to the occasion. He followed Percy Shelley out of the room, took hold of him, and administered a good dose of ether.

Later that evening Polidori jotted in his diary, "June 18—Lord Byron repeated some verses of Coleridge's Christabel, of the witch's breasts; when silence ensued, and Shelley suddenly shrieking and putting his hands to his head, ran out of the room with a candle. Threw water in his face, and after gave him ether. He was looking at Mrs. Shelley and suddenly thought of a woman he had heard of who had eyes instead of nipples, which, taking hold of his mind, horrified him."

As literary history and Mary Shelley's own recollections have always stated, it was during one of those nights in June that *Frankenstein* fully presented itself to her as the story she had been waiting all her life to tell.

"I placed my head on the pillow, I did not sleep . . . My imagination, unbidden, possessed and guided me, gifting the successive images that arose in my mind with a vividness far beyond the usual bounds of reverie," Mary Shelley detailed in her introduction.

FRANKENSTEIN

By the glimmer of the half-extinguished
light I saw the dull yellow eye of the
creature open; it breathed hard, and a
convulsive motion agitated its limbs.
. . . I rushed out of the room.
Page 43

London; Published by H. Colburn and R. Bentley, 1831

Frankenstein observing the first stirring of
his creature; print from the 1831 edition.

She continued, "I saw—with shut eyes, but acute mental vision—I saw the pale student of unhallowed arts kneeling beside the thing he had put together. I saw the hideous phantasm of a man stretched out, and then, at the working of some powerful engine, show signs of life, and stir with an uneasy, half-vital motion."

Victor Frankenstein and his unearthly creature had finally arrived.

Chapter 7

FRANKENSTEIN;
or, The Modern Prometheus

Did I request thee, Maker, from my clay

To mold me man? Did I solicit thee

From darkness to promote me?—

John Milton, *Paradise Lost*

IN THE DAYS FOLLOWING THE START OF BYRON'S GHOST story competition, John Polidori spent a lot of time with the Shelleys, particularly with Mary. His diary entries include short notes about walks he took with Mary, as well as descriptions of the animated dinners and numerous conversations they shared. There is little detail about their creative minds; one entry, from June 28, says only, "all day at Mrs. Shelley's."

What did they speak of during these times together? The contest's participants had agreed not to reveal anything about their stories in progress while the writing was being done. But did Mary mention, even in passing, that she had begun to write? Did she tell Polidori what direction her writing was taking? Did she ask him for any scientific and anatomical information only a doctor might know? She had to have been aware of his feelings for her, but did she, in any way, take advantage of Polidori's infatuation to sharpen her tale? If so, Polidori never made note of this in his diary, or if he did, his sister may have put it in the flames because she thought it was too salacious.

By early June, Claire had told Percy about her pregnancy, though it's unclear when they told Mary. But once Lord Byron became aware of the situation, he was not pleased. He knew Claire would need money to care for the child, and instead of accepting his responsibilities, he seemed offended and affronted. Observing this situation, Shelley had legal papers drafted that allocated a portion of his inheritance to Claire and her child.

Byron had never been entirely fond of Claire, and now he felt conned by her. He went so far as to doubt the child's paternity: "I never loved her nor pretended to love her, but a man is a man, and if a girl of eighteen comes prancing to you at all hours there is but one way—the suite of all this is that she was with child—and returned to England to assist in peopling that desolate island," he later wrote to his friend Douglas Kincaid. "Whether this impregnation took place before I left England or since I do not know; the [carnal] connection had commenced previously to my setting out—but by or about this time she is about to produce—the next question is, is the brat mine? I have reasons to think so, for I knew as much as one can know about such a thing—that she had not lived with S. during the time of our acquaintance—and that she had a good deal of that same with me."

With summer quickly coming to an end, Mary, Percy, and Claire left the shores of Lake Geneva on August 29. Claire and Lord Byron, though mostly Byron, had decided that Claire's child would remain with one of its parents until at least the age of seven. And in order not to taint anyone's reputation further, Claire would be referred to as the child's aunt, which would allow her to see and care for the baby as she wished without inciting malicious rumors.

Almost two weeks later, on September 16, John Polidori also left Villa Diodati. He marked it in his journal simply as "Left Cologny and Lord Byron at six in the morning."

By now Byron had become busy with his own life and had had enough of Polidori's juvenile antics. Byron had wanted a physician who would help restore his mental and physical faculties. Instead, in Polidori he got someone who, despite his age, still acted like a boy and needed near-constant supervision, discipline, and assurance. Byron was unwilling to offer any of these.

A few days after departing, Polidori wrote to his father, Gaetano, to tell him what had happened and to let him know his plans: "I was in agitation for my parting from Lord Byron. We have parted, finding that our tempers did not agree. He proposed it, and it was settled. There was no immediate cause," he explained, "but a continued series of slight quarrels."

Blaming himself for what had occurred, Polidori added, "I am not accustomed to have a master, and therefore my conduct was not free and easy." In the upcoming months, he planned to make his way toward his father's native land, Italy, and there, if opportunities presented themselves, to try to settle as a doctor.

The letter seemed to give Gaetano a certain amount of satisfaction that he had been proven correct and that Polidori had decided to leave "a man so discredited in public opinion." But still, having glimpsed a streak of indecisiveness in his child, he feared that in the months to come, he would see his son become "almost a vagrant."

Byron was thrilled to have gotten rid of Claire's presence and lustful gazes, and he was also happy that Polidori was gone. In a later letter to his sister, Byron described Polidori as merely useless. In the young man's desire to please, he had been more of a hindrance. Still, Byron had to admit that his health had improved in the previous months, but he could not tell if that was due to Polidori's ministrations or simply having been away from the nasty rumors of London.

There were others in Byron's circle who found Polidori distasteful and were relieved to see him leave. John Hobhouse, one of Byron's closest friends, wrote in his diary: "Helped Dr. Polidori to settle his involved accounts with Lord Byron, and took leave of him . . . He is anything but an amiable man, and has a most

unmeasured ambition, as well as inordinate vanity; the true ingredients of misery."

Polidori's own diary entries lapsed as he sallied through Italy
with a lot of hope but very few financial resources. When he landed
in Milan, he stayed for a month and reconnected with friends and
acquaintances whose social calendar included galas, dinners, and
most especially concerts at the famed Scala opera house. Though
he enjoyed those outings and Milan society, his medical career did
not prosper there; actually, there was no indication that anything
remotely resembling a career was about to be established. As his
savings dwindled, he traveled to Pisa, where, under the supervision
of a local doctor, he began to care for Italians as well as English
expatriates living in the area. Three people under his care died due
to mysterious circumstances, and following local investigations,
their deaths were attributed in part to his substandard care and his
general skills as a physician were questioned.

Italy was proving to be no better for him than Switzerland had
been. With his patients dying and finances shoddy, Polidori wrote
to his father again asking for money to return to England. He had
come to believe that what his father had told him about his native
land was all "but too true."

IN PROBABLY THE ONLY BOOK WRITTEN ABOUT POLIDORI, *POOR POLI-
dori: A Critical Biography of the Author of "The Vampyre,"* D. L. MacDonald
described the young man as being "so unsuccessful in everything
he tried to do" that he went about classically "demonstrating the
pattern of compulsive failure that Carl Menninger regards as a
form of chronic suicide." This propensity for failure became even
more obvious when he returned to England.

He had become a doctor at his father's urgings, and now he no longer found the work or the patients appealing. He also became convinced that perhaps he should try something else and began to study law, but soon found that he had no aptitude for it and was bored by it. As he had done while on Lake Geneva, he began to frequent bordellos and houses of gambling, for which he felt a strange and repulsive attraction. Though he was book-smart, Polidori's intelligence did not translate well to a game of poker or a hand of cards, and inevitably he became indebted to loan sharks who were eager to collect on what he owed them. While returning to his rooms following one particularly nasty visit to a gambling hall, an incident occurred that, according MacDonald, caused Polidori to suffer "brain damage."

On September 20, 1817, the *Norfolk Chronicle* and *Norwich Gazette* gave sketchy details of an accident that took place in their area. A few days earlier, on September 14, a "Dr. Polydore" had smashed his gig into a tree. "The night being dark . . . going at a slow rate, he drove against a tree, upset, and broke the gig, and following on his head, a violent concussion of the brain was the consequence . . . He remained for several days in an almost senseless state."

They could not tell if the concussion left any permanent marks on Polidori, but after that, his life unraveled even further.

His true desire had always been to make a mark on the literary world. Following that passion, he continued to write and seek employment from many places, including with John Murray, Byron's publisher and the man who had once considered publishing Polidori's travelogue. But Murray would not answer Polidori's inquiries, and Polidori could not understand why. If he had read an earlier letter of Byron's to Murray, he would have known. After

Byron heard what had occurred in Pisa, he had written to Murray: "I was never more disgusted with any human production than with the eternal nonsense . . . and ill humor and vanity of that young person."

On April Fool's Day of 1819, while still residing in his room at the Covent Garden Chambers, Polidori began to read the day's edition of Henry Colburn's *New Monthly Magazine.* In it he discovered "The Vampyre: A Tale by Lord Byron" and was startled, to say the least.

He was surprised not only because Byron, a man who did not believe in such things, had written a story about vampires, but also because Polidori was the one who actually wrote "The Vampyre."

Next to the story, the magazine had added an addendum noting "the tale which accompanies the latter we also present to our readers, without pledging ourselves positively for its authenticity, is the production of Lord Byron."

How had the magazine gotten their hands on "The Vampyre"? Polidori had not sent it to them for publication. Eventually, it was revealed that someone who knew the whole Lake Geneva party had found a copy of "The Vampyre" and sent it to Colburn, attributing its authorship to Byron.

But before that was known, Polidori wrote to the journal and proclaimed that they had been "led . . . into a mistake in regard to the tale of the Vampyre which is not Lord Byron but was written entirely by me." Even Byron wrote to the magazine: "Damn *The Vampyre*—what do I know of vampires?" He declared to Colburn: "I have seen mentioned a work entitled '*The Vampyre*' with the addition of my name as that of the author—I am not the author and never heard of the work in question until now . . . if the book is clever it would be base to deprive the real writer—whoever he might be—of his honours—and if stupid—I desire the responsibility of nobody's dullness but my own."

The magazine was forced to retract its mistake but did so only partially. They reissued "The Vampyre," this time with Polidori's name, albeit with a line that described it as being a "more extended development" by this author.

Why had such fanfare ensued over "The Vampyre"? Was it simply a matter of mistaken or stolen authorship? While arguably not the best or most terrifying gothic story ever written, in Polidori's hands, the vampire had been transformed from a common bloodsucking winged creature of the night to a slick seducer, albeit still a bloodsucking one, who enraptured his prey not with fear or loathing but with the promise of nightly sensual delights. "The Vampyre" is about Lord Ruthven, a sexy aristocrat whose prowess and libido are similar, not by coincidence, to Lord Byron's. This was also the first English-language account of a half-man and half-bat creature, which captured the public's imagination and popularized the idea of the modern-day vampire. Several editions were printed and reprinted and various very successful stage adaptations took place. Only after the 1893 publication of Bram Stoker's *Dracula* did "The Vampyre" take a backseat.

In August 1821, English journals ran a short article that said a "Melancholy Event" had taken place in their midst. It was learned that John Polidori had, some days earlier and while still residing in his gloomy rooms in London, ingested a large dose of prussic acid and taken his own life. The night before his demise, a servant had gone to his rooms and "found him groaning in the last agonies of death." Concerned, the servant summoned several doctors, but they quickly determined that the poison had already traveled through his body. He died some hours later, and the doctors described the official cause of death as having been not prussic acid, but "the visitation of God."

The obituaries were not kind to Polidori. They declared that "the deceased . . . had for some time accompanied Lord Byron to Italy." They did not mention any details about "The Vampyre" or that he had been one of the youngest graduates of one of the most distinguished medical schools in the British Isles. They did not find it necessary to detail his publications, although they were few, or his thesis on somnambulism. His greatest accomplishment, they felt, was to have played second fiddle to Lord Byron. Polidori knew he would always be remembered for this so-called accomplishment.

Earlier in his life, while still traveling with Byron and keeping a diary, he jotted down an entry after meeting a roomful of people. It encapsulated where he felt he stood in the grand scheme of things: "May 28—Introduced to a room [full of people]—Lord Byron's name was always mentioned, mine, like a star in the halo of the moon, invisible."

Lord Byron felt sorrow for a time after hearing of Polidori's death. He recalled their many conversations, some of them dealing with the possibility of "taking prussic acid, oil of amber, blowing into veins, suffocating by charcoal, and compounding poisons." That Polidori had actually now gone ahead and taken his own life did not surprise Byron; he realized that perhaps "disappointment was the cause of this rash act."

IN THE MONTHS FOLLOWING THE SUMMER OF 1816, DEATH SEEMED TO shadow the members of the Lake Geneva party and all of those who were, in some fashion or another, affiliated with them.

Percy, Mary, their baby, and Claire returned to London during the first week of September. Percy headed to Marlow to find a home for them, while Mary and Claire made a quick dash to Bath, where they planned to sequester Claire so she could wait for the arrival of her baby. Claire wrote to Byron that Bath was "a very fine airy town, built up the sides of hills in high terraces." But even with the beauty that surrounded her, she found it solemn and boring. The Godwins had not been told of her pregnancy.

Even though Byron had made it clear he wanted nothing to do with her, Claire continued to hope for a future with him, writing letters meant to sway him: "You should have a nice house to live in, my nice little girl (I hope it will be a girl) to educate & the friends who could best visit you & we should have nice poems written (to) by you & copied by little me to improve this vile world which always reviles in proportions to its envy." Lord Byron did not reply to any of her letters.

Through it all, Mary continued to write her story.

Her father and stepmother had not changed their position toward Mary and Percy, and worst of all would not allow Fanny to visit them.

Finally, Fanny decided to take control of the direction of her life. In late September 1816, when she climbed aboard a coach bound for Bristol, everyone believed she was on her way to visit her aunts. But Fanny had already made up her mind about how this trip would progress and also how it was going to end. On reaching her destination, she wrote two letters, one for her sister Mary, and one for the only father she had ever known. The letters said she wished to "depart immediately to the spot from which [she] hope[d] never to remove." Then, taking a deep swig from a bottle of laudanum, Fanny Imlay committed suicide, succeeding where her mother had failed twice.

These actions were those of a young woman who not only believed she was trapped in a bad situation, but also saw herself as a burden to those she loved and who loved her. Her conundrum was explained in a note she left on the bedside table: "I have long determined that the best thing I could do was to put an end to the existence of a being whose birth was unfortunate, and whose life has only been a series of pain to those persons who have hurt their health in endeavoring to promote her welfare. Perhaps to hear of my death will give you pain, but you will soon have the blessing of forgetting that such a creature ever existed."

When the letters arrived, William Godwin and Percy Shelley both set out for Bristol, but they arrived too late. By then, *The Cambrian*, the newspaper of the area, had already published the "melancholy discovery" of Fanny's body. The script told of how "a most respectable-looking female arrived in the Mackworth Arms Inn on Wednesday night by the Cambrian coach from Bristol . . . Much agitation was created in the house by her non-appearance yesterday morning and in forcing the chamber door, she was found a corpse, with the remains of a bottle of laudanum on the table, and a note." At the time of her death she had been wearing a "blue-striped skirt with a white body . . . & appeared 23 years of age, with long brown hair, dark complexion."

Godwin and Shelley returned to London with the news. Mary was eager to go to Bristol and see her sister, but her father warned her not to. He did not want their extended family and acquaintances to know that Fanny had taken her own life. Instead he wanted them to believe she had been struck down by a fatal illness while on her journey. Godwin instructed Mary to let things be.

It wasn't long before the Godwins began to place blame on Mary and Percy for Fanny's suicide. "From the fatal day of Mary's elopement, Fanny's mind had been unsettled," Godwin wrote to a friend. "Her duty was with us; but I am afraid her affections were with her." They felt that Fanny's split loyalties between her sister and Shelley and her stepfamily had caused Fanny to break apart.

Mrs. Godwin, as always, was more vocal. In her opinion it was not so much broken loyalties that had destroyed Fanny, but rather a broken heart. Years after the event, the Godwins wrote to Maria Gisborne, a friend of the family. In the letter they detailed how in love Fanny had been with Percy, how broken she had become when she learned her affections were not reciprocated. "Mr. G. told me that the three girls were all equally in love with ———," wrote Maria Gisborne, "and that the oldest put an end to her existence owing to the preference given to the younger sister."

As those rumors spread, Fanny became pitied by some, mocked by others, and referred to as "a very plain girl and odd in her manners and opinions," though she had been "upright and generous. She was pitied and respected."

Blaming Mary and Percy allowed the Godwins to distance themselves from the act and the situation. They did not have to think themselves partly responsible, as they were, for what Fanny had endured.

Mary, Percy, and Claire secluded themselves to mourn Fanny, a sense of melancholy settling upon Mary's household. They must have known that while they weren't to blame for Fanny's suicide, they weren't completely innocent either. They had expressed a certain degree of callousness toward Fanny, and Mary must have felt a deeper sense of guilt than the rest.

More shocking news came when they learned of another death. And while it was tragic, they had to admit (although never aloud) that this passing came with a certain measure of advantage, if one could call it that. On December 10, the body of Harriet Westbrook, Percy Shelley's legal wife, was found in tattered clothes, bloated and floating in the Serpentine River. No one recognized her right away, and only later was the body identified. She had also committed suicide, and Shelley received the news from his friend Thomas Hookman Jr.

Harriet Westbrook had also left a note, but unlike Fanny's, this one was more bitter and meant to hurt, placing guilt and blame for her demise directly on Percy Shelley and his deeds: "My dear Besshee, let me conjure you by the remembrance of our days of happiness to grant my last wish. Do not take your innocent child from Eliza [Harriet's sister] who has been more than I have, who has watched over her with such increasing care. Do not refuse my last request, I never could refuse you and if you had never left me I might have lived, but as it is I freely forgive you and may you enjoy the happiness which you have deprived me of."

At the time of her death, Harriet was carrying a child whose paternity was unknown. Rumors soon began to circulate that not long after Shelley left her, she became involved with several men at once and the child she had been carrying could have belonged to any one of them.

Shelley quickly went to London, intent on taking custody of

his two children, Ianthe and Charles. He felt that as their father, he not only had a legal right to them but also a moral one. They would live in his household, which Mary supported. But Shelley had not considered that Harriet's family might disagree with him, and that is precisely what they did, taking Shelley to court and suing him for custody, which they won.

While in London, he finally realized what he had done to Harriet and how badly he had handled the events leading up to his departure from the marriage. Harriet and the children had been forced to return to her father's home, but they were driven away for monetary reasons. Penniless and virtually without options, Harriet could only care for herself and the children by using her body as a prostitute. She had lived for a time with a man named John Smith, but he had also abandoned her, causing her further emotional upset. Unable to withstand the humiliating desertion of another man, she had jumped into the river. Even worse, in a letter, Shelley accused Harriet's sister, Eliza, of murder, saying it was her desire for Mr. Westbrook's money that had killed her sister, Harriet.

Of course, the last comment was not true. Eliza was overbearing and always in the way, but she had not killed her sister any more than Percy had. But just as the Godwins had done with Fanny's death, placing blame on Eliza for Harriet's passing allowed him to think of himself as blameless.

On December 30, 1816, just weeks after Harriet's death, Percy Shelley and Mary Godwin became husband and wife in the halls of St. Mildred's Church in London. It appears that a very eager Mary pressured Percy to marry her soon after Harriet's passing. Later in life, during her interview with Captain Silsbee, Claire Clairmont reported that Mary had threatened to commit suicide if Shelley did not marry her. "Mary sat at the end of room . . . When Shelley objected he cd not marry against his principles . . . She advanced put her hand on his shoulder said . . . if you don't marry me I'll do as Harriet did." Silsbee's notes continued: "S. turned very pale Miss C. says after these 2 deaths/suicides one summer one in Dec. Shelley was never the same."

William and Mrs. Godwin acted as witnesses. Godwin was pleased that his daughter had finally married, though he played coy about the marriage and the man she had married. "Her husband is the oldest son of Sir Timothy Shelley . . . ," he wrote to Philipp Huel Godwin, "so that according to the vulgar ideas of the world, she is well married; I have great hopes that the young man will make a good husband. You will wonder how a girl with not a penny of fortune, should meet with so good a match. But such are the ways . . . of this world. For my part, I care little about wealth." In the letter, he clearly lied. Always destitute, he wrote to family, friends, and acquaintances in what he termed "begging letters," trying to find those who could provide him with support.

ON JANUARY 12, 1817, CLAIRE GAVE BIRTH TO A BABY GIRL. MARY, proud of her new status as a married woman, wrote to Lord Byron about the birth and told him about her marriage to Percy. "I took it upon myself the task & pleasure of informing you that Clara was safely delivered of a little girl yesterday morning . . . at four . . . Another incident has also occurred which will surprise you, perhaps. It is a little piece of egotism on me to mention it—but it allows me to sign myself—in assuring you of my esteem & sincere friendship, Mary W. Shelley."

Although the last months of 1816 had been full of sorrow and personal calamities, Mary continued to write her book. By the time spring arrived, she already had a first draft, and by the middle of May 1817, she felt the hardest job had been done. In a diary entry of Wednesday, May 14, 1817, she wrote, "Read Pliny and Clarke—S. reads Hist of Fr. Rev. and corrects F. Write Preface—Finis." It was now time to find a publisher.

As summer neared, work began in earnest to find a home for *Frankenstein; or, The Modern Prometheus.* This fictional tale of a zealous scientist raiding graveyards in order to build a creature did not immediately find a publisher, and by the middle of June, rejections began to trickle in. John Murray, the well-known publisher and friend of Byron, was the first to say no. Although Mary always maintained that it had been more her husband's idea to seek publication, she had to admit that those rejections stung. The whole ordeal put Mary in a bad mood.

At the end of summer, the manuscript was finally accepted by Lackington, Hughes, Harding, Mavor and James, and was published in January 1818. The firm was neither a newly established nor a disreputable one; Shelley had done some deals with them previously. As the year of 1818 began, copies of the book were sent to family and friends, and critiques were eagerly awaited.

According to the *Quarterly Review*, *Frankenstein* "inculculates no lesson of conduct, manners, or morality . . . it fatigues the feelings without interesting the understanding; it gratuitously harasses the heart, and only adds to the store, already too great, of painful sensation . . . the reader [is left,] after a struggle between laughter and loathing, in doubt whether the head or the heart of the author be most diseased." Other journals were quick to point out the flaws,

the abundance of gory details, and what they felt was the somewhat amateurish writing. Some were slightly gentler, lauding the
author's imagination and the style of writing.

There was also the mystery of authorship, because the book
had been published anonymously. This was not an uncommon occurrence in the 1800s, particularly when a woman was the author.
Given that Mary was writing about subjects usually reserved for
scientific men, it had been decided to leave the authorship of *Frankenstein* blank. Thus, readers were left to wonder and guess at the
mind lurking behind those pages. Earlier in his career William
Godwin published a novel, *St. Leon*, that spoke to similar themes,
and many critics now questioned if he was the author of *Frankenstein*. Others felt that not Godwin but his onetime disciple and now
son-in-law, Percy Shelley, had.

Shelley was known to have an affinity for such subjects, and
his published works—the direction they took and the way he expressed those ideas—made the comparisons to *Frankenstein* obvious.

In 1815 Percy Shelley had published a poem titled "Alastor."
In it, a young and idealistic narrator seeking wisdom, truth, and
knowledge questioned Nature in order to find the answers to his
ever-increasing inquiries and musings. Like Victor Frankenstein,
the poem's narrator has been on a journey in search of inner and
outer guidance, but has discovered nothing. Such a search had
brought the "Alastor" narrator, just like Frankenstein, to question
death, both figuratively and literally:

> *And my heart ever gazes on the depth*
> *Of thy deep mysteries. I have made my bed*
> *In charnels and coffins, where black death*

Keeps record of the trophies won from thee.
Hoping to still the obstinate questioning
Of thee and thine, by forcing some lone ghost
Thy messenger, to render up the tale of what we are.

Victor Frankenstein's desire to make the creature stems, in part, from a desire to learn about that evil snatcher he knows as death. Just like the narrator in Shelley's poem, Frankenstein raids coffins and cemeteries, as well as death houses. This is done not only for practical reasons—he needs body parts to stitch together—but because he yearns, through the process of decomposition, to learn what death, and life, is all about. "I collected bones from charnel houses . . . the dissecting room and the slaughter-houses furnished many of my material," Frankenstein says, using almost the same words Percy Shelley had. Frankenstein's words are a combination of Shelley's childhood antics, a young adult's poetic musings, and the ideas spewed out by the natural philosophers of the time.

Although the similarities between Victor Frankenstein, the real-life scientists, and Percy Shelley were obvious—Mary even chose the name Victor as an homage to Percy, who had used that name in his youth because he felt it showed power and strength—Mary always maintained that no one, least of all her husband, had directly influenced her work. "I certainly did not owe the suggestion of one incident nor scarcely of one train of feeling, to my husband," she adamantly wrote in the introduction to the 1831 text. "And yet for his incitement, it would never have taken the form in which it was presented to the world."

What incitement was she referring to?

Most critics were also not only quick to show the similarities between *Frankenstein* and the works of natural philosophers like Humphry Davy, but also to point out the parallels to myths and legends of eras past, most especially because of the novel's second title: *The Modern Prometheus.*

The legend of the Titan Prometheus was well known to Mary, as it was to Percy, who, during the time of Mary's writing, was also composing a work based on the Greek mythological figure. Percy's efforts culminated in 1820 with the publication of *Prometheus Unbound,* a four-act stage play.

The Greek legend told of Prometheus's efforts to help humanity by pitting himself against the powers of Zeus. Prometheus had tried to steal fire from Zeus so he could bequeath it to humans. But when Zeus learned he was being deceived, he became enraged and punished Prometheus by shackling him to a rock for eternity. To inflict further punishment, Zeus sent a bird, most often described as an eagle, to devour Prometheus's liver, which regenerated day after day and was then again eaten by the same bird. In one famous version of the myth, it was Hercules who saved Prometheus. While traveling in the area, Hercules came across the Titan, unbound him, and killed the bird.

In the original myth, Prometheus, while troublesome, was seen as a source of strength. In later retellings, Prometheus morphed into a rebel, a youthful challenger to Zeus's almighty and omnipotent powers. It was his assault on something or someone bigger and more omniscient than himself that caused his downfall. In even further retellings, Prometheus not only stole fire, a symbol of knowledge and power, but was also accused of creating man from clay.

The comparison between Prometheus and Mary Shelley's Victor Frankenstein was not that hard to see. Just like Prometheus,

Frankenstein also tried to steal what was most sacred from God: the ability to create man and control life. By questioning those powers that were beyond his control and level of understanding, he caused his own downfall.

Perhaps, some critics felt, the author had a great understanding of the philosophical debates and moral implications that arose from discussions about nature, the quest for knowledge and power, man versus God, and man's ability to create another entity without God's help. After all, at its most basic level, the book also spoke to society's fear that scientists were delving into regions unknown to them and would, by their own efforts and uncontrolled desires, create their own individual creatures to unleash on the world. The book was, some realized, a scathing critique of society, science, and religion. Given that, it was not surprising that people thought William Godwin had written it.

The Godwin family did not bother to correct reviewers. They were glad the author's true identity had not been discovered.

In June 1818, when William Godwin learned that *Frankenstein* would be critiqued in the *Quarterly Review,* he anxiously awaited the article's publication. But nothing came of it. "The article is very innocent," he wrote to Mary. "They say that the man who has written the book is a *man of talents;* But that he employs his powers in a way disagreeable to them."

None of the reviewers ever considered that a woman might have created the text. No one saw the subtle but evident signs: that the story took place within a nine-month period, a typical gestation period for a pregnancy. They also did not note the attachment disorders both Victor and the creature seemed to suffer from, nor that a woman would know that such disorders stem from the lack of a mother figure. The critics also missed the judgments directed toward a society that thought so little of women, which were plainly evident in the way the female characters were depicted. No one paid attention to or commented on those matters.

Claire Clairmont seemed to be the only one who expressed the significance of a woman's having written such a text: "Mary has just published her first work a novel called *Frankenstein or, The Modern Prometheus*," she wrote Lord Byron in January 1818. "It is a most wonderful performance full of genius & the fiction is of so continued and extraordinary a kind as no one would imagine (to belong to) could have been written by so young a person. I am delighted & whatever private feelings of envy I may have at not being able to do so well myself yet all yields when I consider that she is a woman & will prove in time an ornament to us & an augment in our form."

As the weeks passed, the critics continued to speculate that either Godwin or Shelley had written *Frankenstein*. In a way, they were right and wrong at the same time. By birth and marriage, Mary was both a Godwin and a Shelley.

Chapter 8

THE ANATOMY ACT

He who fights with monsters might take care lest he thereby become a monster. And if you gaze for long into an abyss, the abyss gazes also into you.

FRIEDRICH NIETZSCHE, *BEYOND GOOD AND EVIL*

ON AUGUST 27, 1818, ALEXANDER LOVE, A GENTLEMAN in his late seventies living on the outskirts of Glasgow, awoke shortly after midnight, eager to begin another day of work. He also woke up his young grandson, a ten-year-old boy, and following a light breakfast and those personal ministrations an aging body and a young one required, they left their home and set out for their destination. It was by now nearing dawn, one of those balmy and windless early summer mornings where the humidity thickened the air and stuck to one's skin.

Much like his surname indicated, Mr. Love was a kindly, "inoffensive," and "industrious" sort of fellow, and due to his personal disposition and his relatively advanced age, he neither sought nor ignited trouble with anyone who crossed his path. But trouble was about to find Mr. Love that morning.

As grandfather and grandson slowly walked the narrow and darkened streets, they saw thirty-year-old Matthew Clydesdale nearing them. Clydesdale was a weaver, but he was also a mean drunk, and when Love and Clydesdale met, the younger of the two was thoroughly and viciously inebriated. No one knows for certain what sparked their argument or what instigated the outburst, nor, for that matter, what was said between the two. But upon seeing the old man and the young boy, Clydesdale became so enraged he assaulted them both, brutally killing Mr. Love and severely injuring the boy.

Some newspapers reported that Clydesdale had used a pick-axe, while others said his hands had done the deed. Regardless of what method he had used, following the murder Clydesdale left the area and the two victims—one still breathing—by the wayside and walked toward his home. He arrived there with wet blood on his clothing, and when asked what had happened, he said some drunken fellows had attacked him and tried to rob him, and he had had to retaliate.

He didn't try to deny that he was a drunk and even acknowledged the conditions that overtook him when he drank too much. It was said that "there was something in the temper of Clydesdale when he drank spirituous liquors that his most intimate acquaintances never could fathom; and it was thought that he had momentary fits of derangements when in liquor." Thus it was believed he had committed the murder and assault precisely because of the excessive amount of booze he had ingested.

Matthew Clydesdale was charged with murder and brought to trial on October 3, 1818, in the old High Court of Justiciary in Glasgow, before a Lord Gillies. For better or worse, his counsel turned out to be a young man named William Taylor, a rookie at his job, who had just passed his school tests and the Scottish bar exam that allowed him to practice law. Taylor had learned all he could about the case and intended to use every bit of information he had to get his client off. A lanky, skinny, and somber man, Taylor was, in the eyes of many, "cadaverous-looking," given his paleness and his refusal to ever crack a smile. Peter Mackenzie, who years later wrote a detailed, albeit somewhat salacious, account of the events that transpired, described him as the very "picture of a potato boggle, fitted to scare away the crows."

But while his looks may have been lacking, his powers of reason and persuasion were impressive. To a packed courtroom, he described what had occurred in grave but clear details, flinging his arms about for emphasis, intonating the highs and the lows, delivering for hours a sermon in a screeching timbre that resembled the cries of those angry crows he had been likened to. When he was done, dampened and somewhat disheveled, the audience applauded with abandon, but it wasn't clear if that was because they had been captivated by his account or because he was finally done.

Lord Gillies was equally and justly impressed by Taylor's speech, admiring the "uncommon style" and his eloquence. Still, despite all the rhetoric, he was not convinced of Clydesdale's innocence, and neither was the jury. Following a brief respite, Clydesdale was found guilty of murder, his sentence being death by hanging on November 4, 1818, between the hours of two and four P.M., as tradition demanded. Not only that, Lords Gillies and Succoth "decerned and adjudged that he shall be fed on bread and water only, till the day of his execution and that his body, after being so executed, shall be delivered up by the Magistrates of Glasgow, or the officers, to Dr. James Jeffrey, Professor of Anatomy in the University of Glasgow, there to be publicly dissected and anatomized."

"Having shed man's blood," the hanging judge added, "by man let his blood be shed."

BY THE TIME *FRANKENSTEIN; OR, THE MODERN PROMETHEUS* WAS PUBlished in early 1818, the study of anatomy across Europe was at an all-time high, and viable cadavers were becoming harder to find and increasingly valuable. The immoral act of unearthing and stealing corpses had become common practice, so much so that funerary rites that had been practiced for centuries had to be adjusted to secure eternal rest for the dearly departed. New apparatuses were erected by those who could afford them; called "coffin-collars," they were cagelike iron structures that straddled the coffins and could be unlocked only by thick iron keys held by the mourners. These cages appeared to prevent body snatchers from getting their hands on the dead body, but resourceful thieves found other ways in. They dug tunnels from the sides, putting themselves in a position that allowed them to hit the coffins at the precise point where the head rested. From there, they cracked through the wood and dragged the body out by the head or the neck. Naturally, this process took longer and increased the risk of being discovered. Regardless of that, they continued on.

Cemeteries hired guards to protect the premises, and groups came together on their own as a sort of vigilante movement. But the height of the body-snatching business occurred during the winter, and those watchers soon realized they needed places not only to hide from the prospective thieves but also to protect themselves from the weather. Small round-shaped structures began to appear in graveyards across England, Scotland, and Ireland. Some were strips of wood nailed together, with a small porthole for viewing and in which a musket could be rested; there was also enough space to lay out a blanket. But others were more ornate: large, bleak, circular fortresses equipped with small working fireplaces; windows surrounding the structure; and space for a reasonably comfortable cot, blanket, and seat, all large enough for a group of watchers.

Some people also put up what they called mort houses. These impenetrable houses were protected by iron bars and offered just a single way in; the coffins rested inside until the bodies decomposed, and then they were buried in the earth. The doors and locks were so secure that grave robbers could not get past them, and if they tried, sophisticated booby traps were set off. Some people took the whole booby-trap business to a vicious extreme, lining the corpses and coffins with quicklime, so that a robber would find a nasty surprise if he cracked open the casket. But instead of quicklime, others used gunpowder, which went off as soon as the seal was broken. The booms were often heard in the nearby villages, signaling that someone had just joined the same people he had tried to disinter.

Attendance at anatomy classes continued to rise, with the attendees demanding that they also be given the opportunity to dissect the corpses. And they were all still particularly interested in the theory of galvanic electricity.

Giovanni Aldini—following his ill-fated attempt to resuscitate the corpse of George Foster—was no longer dabbling in such things, though he still believed he was capable of bestowing life-giving properties. When he returned to Italy, he decided that instead of reanimating the dead with a bolt of electricity, he would do something to preserve life from the damaging effects of fire by building a kind of fireproof armor.

But others continued their work on cadavers, seeing Aldini's experiments as just the beginning of work with galvanic electricity. New laws had to be implemented to stop the more grotesque experiments. In France, for example, an edict had to be passed against overusing cadavers because these so-called galvanic experiments had become spectacles for entertainment and simply a way to shock people.

This is where Matthew Clydesdale's fate was taking him, or at least his body. The deeds that were about to be performed on his body would bring about such infamy, they would be considered even worse than the act he committed himself.

When the courtroom audience learned his sentence would include dissection, "a deep shudder fell" over them, as it did over Matthew Clydesdale. Along with him there had been three other convicts tried that day—two were given pardons and another, a young boy named Ross, who had been charged with robbery, was also sentenced to die alongside Clydesdale. Together they were transferred to the old jail of Glasgow, which "was constantly guarded long before and after that period, with a strong picquet of thirty or forty soldiers."

Though the sentence had startled Clydesdale, he was eventually overwhelmed by remorse and realized that nothing more could be done for him. In order to find solace for his soul and peace in what was left of his days, he sought God's help, and in this case it came in the shape of Dr. John Lockhart, the reverend of Blackfriars Church in Glasgow. Speaking to him, Clydesdale became very sorrowful for his actions.

Giving up drink and finding religion caused Clydesdale to mollify his temper and to ingratiate himself to the prison's longtime governor, John M'Gregor. Despite being "a powerful, strong-built man," M'Gregor had a touch of tenderness in him, and Clydesdale found that part. Clydesdale had become such a well-behaved and model prisoner that a few days prior to his death, M'Gregor entered his cell carrying a bottle of porter for him. Maybe M'Gregor saw this as nothing more than an act of generosity, but he should have known better.

On seeing the amber elixir in its shiny bottle, Clydesdale perked up. He chugged down most of it in one gulp, then politely asked if he could keep the rest, including the bottle. Though it was against the judges' ruling—he could only have bread and water—M'Gregor left the bottle and locked the door behind him. The prisoner was well chained, the doors were securely locked, and the prison was guarded by dozens of heavily armed men; where could Clydesdale go? What could he do? M'Gregor entered his own rooms and quickly fell asleep.

When he woke up the next morning and made his way to Clydesdale's cell, he found a scene so terrifying in its brutality and bloodiness, it momentarily stunned him. Sometime during the night, Clydesdale had broken the liquor bottle—shards of it were strewn across the concrete floor—and, with a chip of it, slashed his wrists and his throat. When M'Gregor found him, he was barely alive, warm blood gushing out from the severe wounds he had inflicted on himself.

Clydesdale's reaction to a sentence of dissection was not uncommon. Prisoners almost always preferred to die by their own hand than the noose of the hangman, especially if dissection was to follow. Just the idea of the anatomist's knife was enough to send a man into a state of panic, so much so that suicide seemed like the better option.

M'Gregor immediately recognized that he had fed the prisoner liquor and could stand trial and be found guilty. He asked for help from a Dr. Corkindale, who often performed surgeries at the prison, as well as a few others who were well regarded in the area. Together these men were able to stop Clydesdale's bleeding and patch him up so he could be led to the gallows and be hanged.

As the date neared, thousands of people gathered before the scaffolds. There were so many people that the execution site had to be heavily guarded by the Fortieth Regiment of Foot. Ready to do his job was Thomas Young, the infamous hangman of Glasgow who was said to have sent so many men to their maker, he had now perfected the art of the drop.

When Clydesdale and Ross arrived, a shout arose from the

crowd. They wore the white gloves and white caps drawn over the ears that were the typical hanging outfit for such events. Facing the crowds, the nooses around their necks were tweaked to ensure a somewhat painless and speedy send-off. Thomas Young might have been good at his job, but he had miscalculated the measurements for Ross, and the young man struggled and flopped about like a carp for some minutes before he passed; the crowd, of course, cheered him on. Clydesdale, on the other hand, died almost immediately.

As the law demanded, the two men were left to hang for some time, until a familiar white horse and cart that had carried away executed criminals in the past drew near the gallows. People watched from rooftops, balconies, steps, windows, lodges, and every other possible empty space as the body of Matthew Clydesdale was carted off to the college. So many people had left their homes to see this that the lawmakers worried someone would try to inflict more harm on the corpse. To prevent this from happening, more officers and the military were hired to accompany the body to its final destination. Still, a caravan of eager spectators followed the horse-drawn cart to the college, hoping to crash its gates and enter to view the dissections. But when the body was inside the building, the gates were quickly shut to prevent further troubles.

Inside the anatomy theater, students and others who had managed to find their way in shuffled by as the two most important professors attending the session entered: Dr. Andrew Ure, who taught chemistry, and Professor James Jeffrey, from the Faculty of Medicine, who by then was well on his way to a record "fifty-eight year occupancy of the chair at Glasgow." Both men were familiar to the Glasgow community, not only for their medical reputation but in Ure's case, for a somewhat sordid personal life.

Ure was extremely opinionated and had developed a combative relationship with a fellow medical man, Granville Sharp Pattison, who had irked him from the start. Pattison, who was thirteen years younger, annoyed Ure because of his methods of dealing with cadavers. In one instant Ure had "complained to his managers about the smell and nauseating remnants from the anatomy lectures." Those complaints and arguments might have seemed like the petty bickering of two ambitious scientists, but they soon turned personal when Pattison was alleged to be having an affair with Ure's wife.

In 1807 Ure had married a young woman named Catherine, who had borne him three children, two sons and a daughter. Their life together had seemed ideal: a growing family as well as a growing reputation that had often allowed Ure to travel across Europe and meet eminent scholars and scientists working in fields like his own. But in 1818, he learned that his wife was pregnant again. A happy occasion turned devastating when Catherine admitted that Ure was not the father of the child and Pattison was.

Ure filed for divorce, claiming his wife had committed "adulterous intercourse with a reputed paramour." The paramour was Pattison, who, aware of the hours Ure kept, had visited the house when Catherine was alone. Servants had become suspicious of all the time the two spent together and began talking.

By August 1818, Mrs. Ure was unable to hide her pregnancy, and Ure took her to a "boarding house" in the country, where he left her alone. In December, she gave birth to a daughter.

Strangely enough, Pattison didn't say much about the affair. He tried to prove it was the work of a melodramatic woman who yearned for her husband's attention, and of a husband who wanted out of a marriage to a woman who was "weak, compliant, opportunistic." There was no evidence of the affair between Catherine and Pattison, nor that he was the father of the new child.

The Glasgow community loved all of this gossip, though Ure tried his best to distance himself from it. He wanted to be recognized for his work and not his personal life. He hoped that his anatomy of Matthew Clydesdale would do that for him.

Among those in the auditorium was Peter Mackenzie, who later gave details of what occurred. As Clydesdale's body was brought inside, a small fire was lit in the middle of the room, and as the white garments that covered the deceased were removed, bells began to toll outside the college.

At this point Mackenzie's account steered clear of the more gruesome details, as the galvanic instruments were applied in, as he said, "a few operations . . . swiftly on, which really we cannot describe." He may have found it difficult to write about this, but others did not, including Ure, who, on December 10, 1818, read a lecture at the Glasgow Literary Society titled "An Account of Some Experiments Made on the Body of a Criminal Immediately After Execution." "A large incision was made into the nape of the neck, close below the occiput . . . A profusion of liquid blood gushed from the wound, inundating the floor," his report later read. "A considerable incision was at the same time made in the left hip, through the great gluteal muscle, so as to bring the sciatic nerve into sight, and a small cut was made into the heel."

When Ure pulled out what he referred to as a "minor voltaic battery," the audience members began to squirm uncomfortably in their seats as "the pointed rod connected with one end of the battery was . . . placed in contact with the spinal marrow, while the other rod applied to the sciatic nerve." When the battery was turned on, "every muscle of the body was immediately agitated with convulsions at each new renewal of the electric contact," so much so that it looked as if the body were "shuddering from cold."

As the doctors moved to the lower limbs, "the leg was thrown out with such violence as to nearly overturn one of the assistants, who in vain attempted to prevent" the body from injuring those nearby.

The audience may have been startled but Ure argued, "The success of it was truly wonderful. Full, nay, laborious breathing, instantly commenced. The chest heaved and fell; the belly was protruded and again collapsed with the relaxing and retiring diaphragm . . . Every muscle in his countenance was simultaneously thrown into fearful action; rage, horror, despair, anguish, and ghastly smiles united their hideous expression in the murderer's face."

The last experiment involved placing the electric power near the spinal cord, where it touched the nerve directly connected to the elbow. This caused Clydesdale's fingers to move "nimbly, like those of a violin performer." Then the fingers seemed to stretch, to extend, so much so that "he seemed to point to the different spectators, some of whom thought he had come to life."

Peter Mackenzie agreed and told his readers that Clydesdale had opened his eyes and stood watching what was happening to him, astonished at being reawakened. He even said that Clydesdale got up from the table for a few moments, which was startling because that meant his neck had not been broken during the hanging. Some members of the crowds were so overcome with terror that they fled, while others fainted.

According to Mackenzie, when Dr. James Jeffrey saw the

corpse rise from the table and fully realized what he had done, he also knew what had to be done next. In a move that very much resembled Victor Frankenstein's, Dr. Jeffrey decided that the newly reborn Clydesdale needed to go back to the land of the dead.

"Dr. Jeffrey pulled out his unerring lancet," Mackenzie wrote in concluding his account, "and plunged it into the jugular vein of the culprit, who instantly fell down upon the floor like a slaughtered ox on the blow of the butcher!"

Many other accounts described what happened that day, but Mackenzie's has lasted and captured the fancy of the readers because he managed to turn a real-life medical experiment into one of those tales that were overwrought with melodrama and so popular at the time. It made for fantastic reading due to its language, its topic, and the thirst people had for such ordeals. But those present at the galvanization of Matthew Clydesdale knew he hadn't come back to life, much less hopped off the table and taken a look around at the audience. This was not because the instructors had failed but simply because he could not, physically, have done so. As the medical transcripts and reports later noted, the act of respiration had not commenced because "the body had been well nigh drained of its blood, and the spinal marrow severely lacerated. No pulsation could be perceived meanwhile at the heart or wrist; but it may be supposed that but for the evacuation of blood, the essential stimulus of that organ this phenomena might have occurred."

Ure suggested that if the blood had not been drained, Clydesdale might have been brought back to life. "This event, however little desirable with a murderer," he said, "and perhaps contrary to law, would yet have been pardonable in one instance, as it would have been highly honorable and useful to science."

～⁂～

ON NOVEMBER 3, 1828, THE STARK CITY OF EDINBURGH WAS AWAK-
ened from its deep sleep by the brutal news that sixteen—possibly
even more—of its inhabitants, living in the West Port district, had
been killed and their bodies sold to the renowned anatomist Dr.
Robert Knox. It was not news that body snatching was occurring,
but now someone had bypassed the usual procedure of digging
up a body that had died due to disease, age, or accident and was
killing people for their corpses. This sent shock waves across the
country and beyond.

The perpetrators, William Burke and William Hare, along
with their female companions Helen M'Dougal and Margaret
Hare, had, over the course of nearly a year, devised a way of smoth-
ering individuals that left virtually no trace of the murder. The
bodies were then sold to the anatomists in the area.

William Burke was from Ireland and arrived in Scotland sometime
in 1817 intent on working on the Union Canal. Married and already
a father to two young children, his wife had refused to follow him
to Scotland, leaving him essentially free to start life not only anew,
but with a new woman—and that's precisely what he did, soon meeting Helen M'Dougal. He was described as being on the short side,
though his body was as compact as a pugilist's. He did not possess that
"sinister expression of countenance" it was believed murderers usually
had. In the later-published *West Port Murders,* a description was given that
summed him up as such: "His face is round, with high check bones,
gray eyes, a good deal sunk in the head, a short snobbish nose, and a
round chin, but altogether of a small cast. His hair and whiskers, which
were of a light sandy colour, comport well with the make of his head."

He was said to have good manners and to walk about so quietly that people were often startled when they noticed he was there.

William Burke met William Hare when they were both laborers on the canal. Unlike Burke, Hare was not married and became involved with Margaret only a few years before the scandal broke out. Margaret had been left a widow for some years and owned a rambling lodging house in the West Port district of Edinburgh, a place called Tanner's Close, described as dark and dank, not to mention lonely enough to allow the eventual tragedies to occur. In time, all four came to live under that roof. Of the two men, William Hare possessed the most "loathsome" appearance and personality. In an article that appeared in *Blackwood's Magazine* in 1829, William Hare was called "the most brutal man," someone who looked like an "ideot." More than that, it was the nastiness people discerned in his appearance that disturbed them, the "sullness and squalor . . . native to the almost deformed face of the leering miscreant . . . So utterly loathsome was the whole look of the reptile."

Neither Burke nor Hare had a criminal record, and though they had held menial jobs before—as everything from laborers to fishmongers to shoe repairers—work itself did not appeal to them. They were always trying to hatch a new moneymaking plan or come up with a get-rich-quick scheme. And as they were spending the few guineas available to them at the local pub, an opportunity presented itself to them.

In November 1827 a tenant of theirs, a pensioner named Donald, died without notice. His death did not cause them any grief, though the old man had died without paying his rent and the men felt wronged, robbed even. They were even more unhappy when they discovered that Donald hadn't left behind anything they could sell.

Of course, they had heard of the resurrectionists, but Burke and Hare were cowards and did not have it in them to rob graves. In the trial that came later, Burke gave the impression that he thought robbing a grave was a much greater offense than murder: "Neither Hare nor myself ever got a body from a church yard," he said. "All we sold were murdered save for the first one . . . who died a natural death in Hare's house."

Seizing an opportunity, they quickly made plans to sell Donald's body to Dr. Alexander Monro of the University of Edinburgh Medical School. As they made their way to the school, they crossed paths with a student of Dr. Robert Knox. The young man told them that Dr. Knox was always looking for cadavers, and his price was much higher than Dr. Monro's. Burke and Hare headed toward Dr. Knox's anatomical laboratory.

Dr. Knox had graduated as a physician from the University of Edinburgh in 1814 and had followed a career that had allowed him to become an assistant surgeon in the navy prior to receiving an appointment at St. Bartholomew's Hospital in London. Though his accomplishments in London had been stacking up, it was to Edinburgh that he returned in the early 1820s, where he quickly married and for a long stretch ran the famous Barclay's Anatomy School in Surgeon's Square.

Dr. Knox had been blessed with a prodigious intellect, which he had displayed early on at the Royal High School and the University of Edinburgh Medical School. But he had also been cursed with an ugly face and an even uglier disposition. To counter that, he always dressed in the latest fashionable styles and effused such knowledge and intellect that people, most especially women, tended to look beyond his spotted face (a leftover gift from a severe bout of

Robert Knox, line engraving. The famed Edinburgh physician's involvement with William Burke and William Hare ruined his career and reputation, though he always denied knowing where the cadavers came from.

measles) to the impeccability of his manners. That impeccability became notorious, so much so that as his reputation as an anatomist grew, so did attendance to his classes. By the late 1820s, nearly five hundred students were attending his lectures—so many that he had to split them into several groups over several days. With such a demand for his lectures, he was always in need of fresh cadavers.

Burke and Hare knocked on the back door of Dr. Knox's laboratory. His assistant Paterson opened the door and let them in with their wooden chest. Paterson looked inside and said the corpse of Donald was acceptable. He would pay the seven pounds the two demanded, more than they had ever earned at any of their jobs. Of course, they drank the money at the local pub. But as they did so, they could not help but reflect on how easily they had gotten the small wad of cash. The whole ordeal, the whole transaction, Burke later said, "made [them] try the murdering of subjects."

Burke and Hare always believed they had good reasons for doing what they were doing. When another occupant in their tenement became ill, they rationalized that they were killing him not so much to sell his body but because he was suffering savagely from his disease. They were providing a kindness toward the failing man, and in turn the selling of his body would provide them with a recompense. Using a method that would become their signature style, they plied the man with drink until he became incoherent, then smothered him by covering his nose and mouth until he suffocated. That style of murder left no visible traces and eventually became known in the common language as Burking.

But while the corpse gave them what they yearned for—a small allowance to cover their needs for a short while—they quickly realized their lodging house was never completely full; at times Burke and Hare and their women were the only occupants. They needed a more reliable source of subjects.

When asked if anyone else knew what they were doing, Burke stated: "None but Hare. We always took care when we were going to commit murder, that no one . . . should be present . . . that no one swear he saw the deed done. The women might suspect what we were about, but we always put them out of the way when we were going to do it."

Both men always maintained that their companions (and everyone else, for that matter) were not aware of their doings, but plenty of reliable information indicates otherwise. The women may not have taken part in the actual killings, but they knew what Burke and Hare did. Not only that, the women often found the victims.

Many of those who died were elderly, prostitutes, indigents, drunks living on the streets—people who in the criminals' minds no one would miss and no one would look for. But those people also yearned for friendship and a place to stay, which set them up to become victims. It was not uncommon for them to be lured to the tenement with the notion of free-flowing drinks, merry conversations, a bed to rest in for the night, and the ability to rinse filthy garments under a trickle of relatively clean water. In this way, for most of 1828, three men, twelve women, and a child came face-to-face with Burke and Hare.

Burking left no evidence that a crime was committed, but Knox and his assistants should have become suspicious when Burke and Hare were visiting the back door a little too often. Also, the corpses may not have displayed any outright evidence that the victims had been killed, but nothing else indicated why they were dead. These corpses were not only fresh but also well preserved. Usually a dead body that had been interred and then disinterred bore the marks of that ordeal. The earth itself left remembrances—even if the body had been buried for only a few hours—and the elements had a way of insidiously infiltrating into the flesh. Aside from that, the corpses were also most often dragged out of their coffins by the necks or their hair, which left some kind of physical trace behind.

Burke and Hare's corpses displayed none of those signs. Sometimes young, and relatively pliable, they appeared to have been going about their business one moment and then were suddenly struck down by a simple act of God. Knox, by reputation and practice, was a skilled physician. Should he have not seen what was going on? Did he notice and choose to keep quiet? Those who knew him adamantly refused to believe he had been aware of the murders, thinking instead that he'd proceeded just like all the other anatomists and had not asked where the corpses were coming from, nor how the resurrectionists had gotten them.

Eventually Knox did know what was happening, particularly when the next few victims trickled in. Burke and Hare's greed reached such a point that they no longer looked toward the indigents of society as their targets, but to anyone who crossed their paths, most especially the prostitutes, who, parading up and down the streets, seemed to taunt the killers with a price advertised on their carefully coifed heads. Burke and Hare, perched on stools inside a bar, their heads bent close together as if discussing the latest political scandal, were noticing the happenings among the crowd inside and out, who would make a fitting subject for the anatomists. This was when they began setting their eyes on the more famous characters and their unraveling took place.

On a fine April morning, Dr. Knox began his customary rituals before his students arrived. A body was set upon the surgical table, one that had been provided by Burke and Hare. By now they had become if not his only suppliers, then his most reliable, and though he had not seen today's corpse, he expected that it would be fresh, plump, and without any advanced case of disease or injury. As Knox and his assistant lifted the sheets to view it, they were rewarded with an amazing if nasty surprise: the body of eighteen-year-old Mary Paterson, who despite her last name was not related to Knox's assistant, lay there in all her nakedness.

This was perhaps the moment when Knox realized something was amiss. Young, beautiful, and extremely healthy, she was a well-known fixture in the streets of Edinburgh not only because of her beauty, but because she was a prostitute. It has been suggested that the men in the room were so moved because many had been her clients.

Mary had been lured to the tenement with the prospect of alcohol and had been smothered to death in the same manner that all the others had been. But she did not suffer the usual fate under Knox's knife—not right away, at least. Reports later revealed that

Knox, his assistants, and those students who had arrived in the laboratory were so taken by her naked flesh that they refused to work on her, and instead of dissecting her, they used her as a model for their artistic sketches, albeit a dead model. Overwhelmed by the beauty of her form, Knox placed Mary's body in an alcohol-filled cask, where it remained preserved for the next three months.

Though people noticed Mary was missing from the streets, her disappearance was blamed on her lifestyle. Perhaps she had moved where opportunity loomed, or, if indeed she had been killed, as some in her line of work were, her body would be discovered some months later. What finally got everyone talking was the disappearance of James Wilson, who was known in the neighborhood as Daft Jamie.

An eighteen-year-old boy who walked with a limp and suffered from mental retardation, Daft Jamie was well known and well liked by the children in the area. He lived with his mother and was on his way to meet her when he came face-to-face with William Burke. Burke said he knew where Daft Jamie's mother was and directed Jamie to Burke's house. Of course Jamie discovered that his mother was not there, but he stayed for a while as the two adults seemed friendly and offered him liquor. Jamie wasn't much of a drinker, so he refused. Burke and Hare wouldn't take no for an answer, so they forcibly tried to make him swallow the spirits, which resulted in a struggle. Burke and Hare must have miscalculated Jamie's physical strength, because he was unusually powerful and strong, and soon he was able to overcome Burke. The two continued to fight, and then Hare intervened; it was a tough fight, but they overwhelmed the boy, smothered him, and sold his body to Dr. Knox. As Jamie's mother and the neighbors searched for him

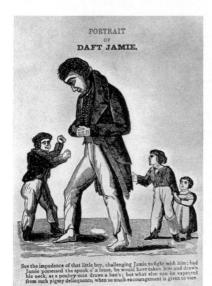

PORTRAIT
OF
DAFT JAMIE.

See the impudence of that little boy, challenging Jamie to fight with him; had Jamie possessed the spunk o' a louse, he would have taken him and drawn his neck, as a poultry-man draws a hen's; but what else can be expected from such pigmy delinquents, when so much encouragement is given to vice.

Portrait of James "Daft Jamie" Wilson, murdered by William Burke and William Hare. Wilson was suffocated by the two killers in William Hare's boardinghouse, and his body later sold to the anatomist Robert Knox for dissection.

and called out his name along the darkening streets, the doctor in his laboratory was dissecting Jamie's body.

Marjory Campbell Docherty had the terrible honor of being Burke and Hare's last victim. She had arrived at the Hares' tenement along with another couple, the Grays, and was killed when the Grays were not present. Neighbors surrounding the dwelling heard a struggle and thought it sounded vaguely peculiar. Then, realizing that it was Halloween, a night reserved for debauchery and wickedness, they had chalked up the noises and screams to a party getting a little out of hand. Following the murder, Docherty's body was hidden beneath a bed. When the sun came up, the Grays asked about Marjory and were told she had left. Ann Gray found that disturbing, because Marjory had not said anything about leaving, and she became even more suspicious when she moved near her bed to remove some undergarments and was blocked by William Burke. It seemed odd that he wouldn't let her retrieve her stockings.

Later that day, Ann was able to return and investigate. The place seemed as dingy and foul as it always did, but she and her husband checked beneath the straw mattress and

Burke and Hare, suffocating Mrs. Docherty. Her body was then sold
to Dr. Knox for dissection. The killers used the method known as
Burking—plying their victim with drink then suffocating her. Mrs.
Docherty was their last victim.

found Marjory's dead body. They quickly left the house intent on
informing the police, but as they rushed out, they bumped squarely
into M'Dougal, who immediately understood what had happened.
Trying to calm them, she promised that they would be part of the
money if they kept quiet. The Grays refused and hurried outside
toward the police.

M'Dougal rushed to tell Burke and Hare what had happened,
and they removed the body from beneath the bed and quickly sold
it to Dr. Knox. The police were later directed toward Dr. Knox's
laboratory, where they found the remnants of Marjory.

Though Burke and Hare had committed numerous murders,
the police found themselves with a dilemma: by the time they arrested the men and their significant others, most of the evidence
had disappeared. The crime spree had occurred over a period of

nearly a year, and the dissections most often took place as soon as the corpses arrived on the table. Afterward the doctors usually buried the remains in unmarked graves, or they incinerated them. The police agreed to give William Hare complete immunity if he not only confessed to all their deeds but also testified against Burke. And "from the hour in which he heard that his associate Hare was to be admitted as evidence against him," *The Scotsman* reported, "he [Burke] abandoned all hope of acquittal and resigned himself to his awful fate."

When Hare arrived on the stand, he provided detailed accounts of the murders, with one particular difference: he took himself out of the ordeals, placing the blame for all that had occurred on Burke's shoulders. He pointed to the method Burke had employed in choosing his victims. He said that one particular incident involved an elderly woman, whom Burke suffocated on the bed while Hare sat by and watched. If Hare intended to make himself come across as sympathetic, it did not work.

He even said that other people, including Knox's assistant Paterson, were involved. To a crowd that was already inflamed, his words only made things worse, particularly when he described how one body had been disposed of in a box with the doctor's assistance: "He [Burke] went in, and drew the body from under the bed, and the porter put it in; there was some hair hanging out, and the porter put it in, and said it was bad to let it hang out."

Hare went on: "The porter carried it to Surgeon's Square. It was roped . . . I went with the porter, and Burke went for the Doctor's man."

Hare denied knowing where the bodies had come from or that he'd had any part in their killings. As he spoke, the "sinister expression in his look" bothered those in attendance because they knew he was just as guilty as Burke. Particularly offensive to them was the "look of evident satisfaction" he had when he finished his testimony, clearly aware that he had literally gotten away with murder.

The trial took place over the Christmas holiday and lasted only several hours. Not surprisingly, the jury quickly decided that Hare, his wife, and M'Dougal would go free. But Lord Meadowbank sentenced William Burke to hang on January 28, 1829. "You may rest assured that you have no chance of pardon," the lord justice clerk told Burke. "I now solemnly warn you to prepare your mind in the most suitable manner to appear in a very short time before the throne of God."

In a further stroke of what seemed poetic justice, he added that Burke's body would be handed over for dissection to the very same professor Burke and Hare had initially wished to do business with: Alexander Monro. Many saw this as a message that this had now become "murder for hire . . . a new species of assassination."

On the day before Burke's hanging, the skies opened up and a great deluge washed over the city. But the weather didn't seem to matter, because thousands gathered in Lawnmarket to view the preparations for the ordeal. They laughed as the scaffold went up, and reports detail how the crowd's chatter mingled with "the din of the workmen and clinging of the hammers." As the day ended and what natural light there had been during the day diminished to nighttime, the workers lit torches around the area, giving the place a more "lurid glare."

Under normal circumstances, the workmen who raised the scaffold would have done anything to get someone else to take this job, but for the Burke hanging, they fought to perform the honor. When they had the scaffold up, some left the area, intent on returning the next day, but others joined the crowds who had laid claim to empty spaces in dark corners and beneath balconies, wrapped in thin blankets that offered little warmth, determined to be there to watch Burke hang. By morning their space would be reduced by more than half by the rest of the population who would join them.

More than twenty thousand people crowded in Lawnmarket to view the execution on the next day. Numerous constables and law enforcement officers had been hired to watch and patrol the area in case of rioting, but they were not needed. Everyone in the crowd agreed with William Burke's fate, so fights did not break out. If anything, there was an air of gaiety in the area that struck some onlookers as out of place. The erratic voices grew to a fever pitch as Burke was brought onto the scaffold and a noose tightened around his neck. They began to chant, "Burke him, Burke him, Burke him . . ."

EXECUTION OF WILLIAM BURKE
From an Etching by Walter Geikie. Edinburgh, 1829

The execution of William Burke. Burke, along with his associate, William Hare, murdered the poor and indigent who crowded the streets of Edinburgh. They supplied bodies to the various anatomist schools in the area for dissections.

The *Newgate Calendar* reported on the crowd's eagerness to do away with the murderer but also wrote about Burke's state of mind: "On Wednesday the 28th . . .

Burke underwent the last penalty . . . Seats commanding a view of the gallows were let at a large price . . . he was assailed by the hideous yells of public execration. The concluding moments of his existence must have caused him the most acute suffering, for, stung to madness by the horrible shrieks . . . he appeared anxious to hurry the execution . . . as if desirous to escape from that life . . . A short, but apparently severe struggle succeeded."

In detailing the day's events, *The Scotsman* instead chose to depict the audience and its reaction, which bordered on either hilarity or hysterics:

> *The struggle was neither long nor apparently severe; but at every convulsive motion, a loud buzz arose from the multitude, which was several times repeated even after the last agonies of humanity were past. During the time of the wretched man's suspension, not a single indication of pity was observable among the vast crowd—on the contrary, every countenance wore the lively aspect of a gala day, while puns and jokes on the occasion were freely bandied about, and produced bursts of laughter and merriment, which were not confined to the juvenile spectators alone—Burke Hare too! Wash blood from the land! One cheer more! And similar exclamations were repeated in different directions, until the culprit was cut down, about nine o'clock, when one general and tremendous buzz closed the awful exhibition—and then the multitude immediately thereafter began to disperse.*

After the execution, the crowd struggled for the lurid relics of the hangman, trying to get their hands on anything affiliated with Burke, such as pieces of his clothing or remnants from the noose.

Dissection quickly followed at the University of Edinburgh Medical College, at the hands of Dr. Alexander Monro. Every precaution was taken to allow only a limited number of people into the anatomy theater, but the judge's sentence had called for a public dissection, so even nonstudents and non–medical men participated. Months later, a pamphlet titled *The West Port Murders* was published anonymously. The unknown writer said he had been present at the hanging and dissection and felt he had to give it a thorough account. "Every countenance bore an expression of gladness that revenge was so near, and the whole multitude appeared more as if they were waiting to witness some splendid procession or agreeable exhibition." He went on: "At the dissection the quantity of blood that gushed out was enormous, and by the time the lecture was finished, which was not till three o'clock, the area of the class-room had the appearance of a butcher's slaughter-house, from its flowing down and being trodden upon."

Burke's remains were on display for hours to allow people to view them. Thousands streamed by the now-mangled corpse, including several females, which everyone in the auditorium thought peculiar. During the dissection, the skin Dr. Monro had removed from the corpse was stolen. Weeks later, the markets began to sell belts, wallets, and book covers all said to have been made from the tanned skin of William Burke. They fetched a very high price. The structure and mold of his skull was later studied by phrenologists, who believed they could divine an individual's personality, artistic sensibilities, and even murderous inclinations from their bone structure. Unfortunately, Burke's skull turned out to be within the normal range, debunking the phrenologists' theories.

Following Helen M'Dougal's release, she made the terrible mistake of returning to her house, where she was viciously attacked by an angry mob and nearly killed. She was fortunate that her house was located near a police station, and they offered her some protection. She was rumored to have traveled to Australia, where she remained until her death in the late 1860s. William and Margaret Hare returned to Ireland, where it was reported Hare was pushed by some coworkers into a pool of quicklime, which caused him to go blind. He spent the rest of his days as a beggar.

Although Burke and Hare did not implicate Robert Knox, nor was he formally charged for any of the crimes, his reputation suffered. Students stopped attending his lectures, positions disappeared, and appointments he wished for never came to be. He moved to London, where he worked at a cancer institution and eventually died in 1862.

The aftermath of the Burke and Hare episode had a strange effect not only on the laws that were passed afterward, but on the literature that was published. The trial was followed by a rise in crime-driven novels that derived from factual accounts. Even nursery rhymes took a turn for the macabre. A particularly naughty one provided children the chance to skip rope while they chanted the whole event in eight simple lines:

> Up the close and down the stair
> In the house with Burke and Hare.
> Burke's the Butcher, Hare's the Thief,
> Knox, the boy who buys the beef.
> Burke and Hare,
> Fell down the stair,
> With a body in a box,
> Going to Dr. Knox.

But more important, in August 1832, what became known as Mr. Warburton's Anatomy Bill passed through Parliament and became affectionately known as the Anatomy Act. In essence, it would drive the resurrectionists out of business.

From 1829, the year news broke out about the Burke and Hare crimes, until the act passed, the bill went through several revisions, but somehow it was always blocked during the very last stages. What helped it, sadly, was an episode in London that was similar to the Burke and Hare crimes, when it was learned that another gang, which included two killers, Bishop and Williams, nicknamed the London Burkers, had also been murdering individuals in order to sell their bodies to the anatomists. A particular gruesome case was that of an Italian boy, which brought London to its knees.

On November 5, 1831, Bishop and Williams were arrested when they tried to sell the body of a young boy to the porter at

The London Burkers—John Bishop *(left)*, Thomas Williams *(center)*, James May *(right)*—were convicted in 1831 for murdering men, women, and children in London, then selling their bodies to medical schools for dissections. Their nefarious deeds came to light when suspicions were cast on the smothered body of a young immigrant boy they were trying to sell, in what became known as the "Italian Boy" case.

King's College. Suspicious of the body's condition, the porter called on the school's surgeon and anatomist, who performed a detailed examination of the body. The boy was said to have suffered great trauma, including a blow to the head and the removal of all of his teeth, and he also appeared not to have been buried at all. The two men were frequenters of the Fortune of War pub, the notorious haunt of body snatchers in London, where a waiter had noticed one of the two felons washing a collection of human teeth with a pitcher of water. They were arrested and put on trial. Found guilty, they were hanged at Newgate on December 5, 1831, in front of a crowd of nearly thirty thousand.

These two cases pointed to something painfully obvious: not only the dead had trouble resting in peace; the living could not either. The struggle for anatomical material had reached such levels that murder was now a way for some individuals to find viable subjects. Not even two weeks after the Bishop and Williams trial, Henry Warburton introduced the Anatomy Act again. This time the bill passed on May 11, 1832, stating in part that "whereas in order further to supply human bodies for such purposes . . . grievous crimes have been committed, lately murder, for the single object of selling for such purposes the bodies of the persons so murdered: And whereas, therefore, it is highly exponent to give protection, under certain regulations, to the study and practices of anatomy, and to prevent, as far as may be, such great and grievous crimes and murder as aforesaid."

The Anatomy Act appointed three people to inspect all the places and laboratories—private as well as institutionally funded—where anatomizations took place; those inspectors could enter the premises at any time and without any notice.

This would allow them to determine if anatomists were adhering to the new regulations. The act also gave the right to allow a dissection to take place to those who possessed the body legally, unless the deceased had said otherwise and those in charge were aware of those wishes. Corpses could not be removed from their place of death for at least forty-eight hours, and even then, someone in authority had to allow this to occur. The act also made it legal for any "member or fellow of any college of physicians or surgeons" to examine the body of anyone who had died and whose body they had received. They would not be "liable to any prosecution, penalty, forfeiture, or punishment for receiving or having in his possession for anatomical examination, or for the provision of this act." In addition, the anatomists were also responsible for burying the remains of the corpses in consecrated ground.

Not everyone was impressed with the new laws. Thomas Wakley, the founder and editor of the medical journal *The Lancet*, declared in an editorial: "Burke and Hare . . . it is said, are the real authors of the measure, and that which would never have been sanctioned by the deliberate wisdom of Parliament, is about to be exhorted from its fears . . . It required no extraordinary Sagacity to foresee that the worst consequences must inevitably result from the system of traffic between resurrectionists and anatomists, which the executive government has so long suffered to exist. Government is already in a great degree, responsible for the crime which it has fostered by its negligence, and even encouraged by a system of forebearance."

With the bill in place, anatomists and medical men could now find fitting subjects themselves, in death houses, in hospitals, and directly from family members. They would not be punished for carrying on such experiments, however peculiar. As part of the deal, they only had to make certain to find a decent place to bury the bodies after they were done with them. Given these new laws, the resurrectionists were no longer needed. The era of the gentlemen in black was forever over. Or so it seemed.

Chapter 9

A SEA CHANGE

First our pleasures die—and then
Our hopes, and then our fears—and when
They are dead, the debt is due
Dust claims dust—and we die too.

PERCY BYSSHE SHELLEY, *DEATH*

MARY SHELLEY PARTED THE CURTAINS OF HER NAPLES apartment and looked out on the barren and desolate Royal Gardens. It was still winter, and the famed rosebushes had not yet bloomed, though her view beyond the Bay of Naples was as stunning in winter as it was in summer. She looked out to the cresting waters as they slammed against the rocky shores, and from where she stood, she could see Mount Vesuvius standing forlorn, brooding like those infamous Italian men smoking pipes. It was a classic Italian scene, admired and sought out by English expatriates, who came to Naples for its beauty and its climate. "So astonishing and so delicious a spot," the barrister Henry Crabb Robinson would write in his diary some years later on visiting Naples.

But even this vista could not stir Mary's soul.

The writing of *Frankenstein* had been satisfying, albeit draining, but finally it was finished and ready to be published. But before the reactions to Victor Frankenstein and his fiend were gathered, Mary, Percy, and Claire left England, because Percy continued to be plagued by health complaints, and he was always searching for an environment that would help his numerous maladies. (They would try to find such a place nearly a dozen times in the next few years, all futile attempts at finding health and peace of mind for Shelley.) They had arrived in Naples in early December 1818, after several other stays in Italy.

Naples, the ancient city situated on the southwest coast of the Italian peninsula, in the region of Campania, was bordered by two volcanically active areas, Mount Vesuvius and the Phlegraean Fields. Mary was also flanked by two volatile characters, Percy and Claire, who pushed her to display two sides of her own personality—calm and abrupt—at the same time. But for this brief period, they had found in Naples a city with such mild weather that it worked as a balm to soothe their irritated moods.

But as Shelley's condition improved—his days were filled with walks and horseback-riding lessons—Mary's mood declined. She despised the Neapolitans, whom she saw as only a shade better and slightly more alive than the pallid ghostly creatures populating Coleridge's phantom ship. She had tried to move beyond her present state, assisted somewhat by walks with Percy through the ancient monuments scattered throughout the region. They also took boats out on the quiet bay, where the water was "translucent and shiny" and they could enter caverns that could only be explored with lanterns. But none of this seemed to lift the unsettling feeling that had pervaded her person.

Percy, Mary, and Claire settled in the Chiaia district, the most famous and desirable area of Naples. They were joined by a Swiss nursemaid named Elise, who had been hired earlier in the year to care for the children; the ever-present Claire; and Claire's daughter, Allegra. In August 1818, Shelley had also met and hired Paolo Foggi, a young man who had already become indispensable. Shelley liked Paolo's industriousness, the way he hurried to be of service, and his talents as a cook. Shelley also believed that Paolo was a reliable addition to the household and someone who could be counted on. But as the three English visitors settled into their new routine, they began to see signs that Paolo Foggi had a tendency to cheat them. This didn't please them, but considering his more uncompromising skills, they decided to view this as a small defect they were willing to overlook.

As soon as Paolo Foggi made himself a part of the household, he set his sights on Elise, and before long the quiet nursemaid and the charming Italian were engaged in a sexual relationship, which soon resulted in Elise's becoming pregnant. Mary was almost instantly irritated by Elise's condition. She may have come to care for Elise too much—or not enough for Paolo—but arrangements for a marriage were quickly made. The ceremony was a great relief for everyone, but soon after the wedding, the Shelleys dismissed Paolo and Elise from their positions.

Disappointed that this relationship had ended badly, the family believed they would never hear from Paolo and Elise again. But as the couple journeyed north across Italy and into England, they were very busy talking wildly and freely about their former employers, whispering to anyone who would listen about a tale that often wavered between the truth and lies.

On December 27, 1818, the strong wails of a newborn little girl were heard in the Chiaia district. Nearly two months later, Percy Shelley rushed toward the Office of Vital Statistics to have her birth registered. The certificate that was issued stated that the baby's name was Elena Adelaide Shelley. In addition, the certificate also revealed that the child belonged to Mary Shelley, though Mary had not given birth during her time in Naples.

When people heard about the child, they thought Shelley had somehow gotten the baby in a shady black-market adoption scheme to find a child for Mary, who was suffering a bleak depression over the death of their infant daughter, Clara, while they were in Venice. Mary never said this aloud, but there is reason to think that she blamed her husband and stepsister for her baby's death. After all, Claire was the reason Percy had gone off to Venice.

Eager to see Allegra, who was in Venice with Lord Byron, Claire asked her brother-in-law to travel there with her, and he agreed. They departed on August 17, and once there, he wrote to Mary and asked her to follow. On August 31, she left with Clara, who was almost one, and with two-and-a-half-year-old William. When Shelley summoned Mary, he did not consider how harmful a trip like this through the Italian countryside in a stuffy carriage could be for Mary and the children. Clara was already ill, and she became progressively worse over the four-day trip. By the time they had reached their destination, Mary, clutching the child to her breast, knew it was too late.

Now, these months later, was Percy trying to find a new child for his wife in Naples? It's not known if Mary was aware of what he was doing or whether she had even agreed to take this Neapolitan newborn into their household. In any case, Elena Adelaide never became a part of the Shelley family because she died on June 10, 1819. However, for those few months, she lived with a set of foster parents in Naples, whom Paolo Foggi and Elise knew firsthand.

Most people believed Elena Adelaide had been rescued from one of the city's many orphanages, but Paolo and Elise told a different story. Once they reached England, they met with Shelley's friends, the Hoppners, and claimed the child's real parents were not some Italian destitutes but were really Percy Shelley and Claire Clairmont. The adoption story was simply a cover story. Claire had been very ill during December 1818, the same month the baby was born. Elise was working for them during this period and knew Claire had taken to her bed for several days.

After the baby was born, Elise continued, Shelley quickly took her, during the night, to the city's foundling hospital, where she remained for several months, until her death. Elise said she was present during all of these events and that a doctor was called to attend to Claire. Few people had even seen Claire during the months when she was supposedly pregnant because she had been sequestered until the moment of the birth.

On visiting Lord Byron in Ravenna, Shelley learned that this story was being whispered among his acquaintances. People had seen the Shelleys and Claire during their stay in Naples. Those who had met them would have noticed if she had been pregnant. Elise also said that Mary seemed strangely out of sorts while all of this was happening. According to Paolo and Elise, in the months leading up to the baby's birth, Mary had been assaulted by such severe melancholy, given the death of Clara, that Percy and Claire had been able to go about their business unbothered. Eventually, Mary grew tired of what was happening and became aware of her stepsister's pregnancy and of the baby's paternity. But despite her knowledge, she did nothing about it because she was either out-numbered or far too depressed.

When Shelley learned that Paolo and Elise were spreading these rumors, he immediately told Mary, who was horrified. She began a campaign to counteract Paolo and Elise's rumors by dispatching letters to family and friends. In them, she did acknowledge that Claire had been ill at the time of the supposed birth, but she said the sickness was something Claire often suffered from and was not related to a child fathered by Percy or anyone else. She emphasized Claire's shyness and that she was sure Claire would never do any-thing to hurt her. Those lines led those who were familiar with the two women to conclude instead that the scandalous rumor was true. Everybody, including Mary, knew that Claire had never been, nor ever would be, a self-controlled person. People recalled the abandon Claire had displayed in pursuing Lord Byron, a notorious womanizer. And despite the family's best intentions, it had become known that Allegra was Byron's daughter. Those who knew the Godwins also knew Claire had had a crush on Percy.

The salacious scandals that arose from Claire's rumored pregnancy in Naples and Percy's indiscretions occupied conspiracy theorists for decades after all the parties had long since passed. In 1936 all official records pertaining to the child's birth, death, baptism, and the months she survived following her birth were found in the vaults of the Vital Statistics offices. But prior to that, a gentleman of strong New England roots suggested that he knew what had truly occurred in Naples—because he believed that Claire had confessed it all to him.

In the early 1870s, a schooner left the small harbor of Salem, Massachusetts, bound for Europe. It sailed beneath a warm sun and fair winds, an auspicious send-off. Its American captain was Edward Augustus Silsbee. A New Englander born and bred, Silsbee was born in 1820 to a ship-owner father also of New England descent. Though young Silsbee had been reared to join the family business, he also had a keen literary and artistic sensibility that made the Romantic poets especially appealing to him. Throughout his life he collected everything he could about Percy Shelley, whose poetry he felt was the essence of the sublime.

Shortly before leaving Salem, Silsbee learned that the poet's sister-in-law, Claire Clairmont, was still living. Now in her seventies, the white-haired Claire lived in a yellow-speckled apartment in the once-aristocratic Palazzo Orsini, in the heart of Florence. When Silsbee reached Florence in October 1872, he traversed the city's cobblestone streets, setting his sights on the old palace. He liked the feel of the tree-shaded courtyard and enjoyed hiking up the screeching staircase that brought him face-to-face with Claire. Of course, he befriended her almost immediately and visited with her daily for more than a year.

He jotted down notes from their conversations in numerous leather-bound notebooks, most often writing in a hurried, hardly legible penciled scrawl, all interspersed with his views on the countries he visited, the natives he encountered, bits of poetry and narrative lines he heard that much impressed him, lists of the accounts he possessed, questions to ask and those he had already asked, and bank codes. Dozens of these notebooks exist, now housed in the vaults of the Phillips Library at the Peabody Essex Museum in Salem, Massachusetts.

In many of their conversations, Claire obliged Silsbee's curiosity with stories about the two poets in her life who were so different and so alike, going so far as to compare, at a certain point, the thicknesses of Shelley's and Byron's lips: "Shelly," she said, "fine & white too thick, B's curved and [not] thick." Silsbee also knew about the friction between Claire and Mary Shelley, and Claire, at one point, readily admitted that there had been jealousy between the two, mostly because Percy Shelley insisted on spending too much time with her. "Mrs. Shelley jealous of her," Silsbee wrote on a later occasion. "Often S. used to walk with her & pay her attention."

Silsbee suggested that Claire must have shown Shelley more of the attention he craved, physical attention that perhaps Mary was unwilling to display: "Claire more caressing," Silsbee wrote.

"This she says S. liked her." A note on a previous page implied that Mary did not like *"carezze,"* the intimate Italian word that refers to caresses or to the act of touching.

After gaining Claire's confidence, Silsbee asked her for intimate details of what had occurred in Naples. The entry for that particular conversation is relatively short but gets right to the point. He tells of an awkward situation Percy and Claire became involved in and says that indeed, Claire was the person everybody had been talking about for years, the woman who had given birth to Percy's child. Those involved in the predicament—supposedly Percy, Claire, and Mary—promised each other to keep the secret, even though it caused Claire great sorrow.

The lines appear to have been written in great haste, as if the author could not contain his excitement because of what he had just learned and could not be bothered with penmanship, either. But had Silsbee truly learned what happened in Naples? Judging by the briskness of his writing, he believed so. But Silsbee's own methods of recording also must be taken under consideration. He jotted down his notes several hours after speaking with her, not during their conversations. It is also possible that Silsbee misunderstood what Claire said to him. Did he correctly transcribe what Claire had told him or did he write down what he wished she had told him? As it stands, the truth of the Neapolitan child is still a mystery to be solved.

SCANNING THE HORIZON HAD NOW BECOME HABIT FOR MARY SHELley. Whether she did so in Marlow, staring at the droopy willow trees hanging like tentacles over the rooftops, or in Naples, where in the distance the volcano puffed, it didn't matter: her eyes always watched the changing light before her, as she could see her own life change in the various colors reflected in the seasons of nature. In 1822 she found herself staring at another horizon, with the new waves of another sea. This time, the Gulf of La Spezia, on the upper northwest coast of Italy, stretched before her, the blowing of the Mediterranean winds bringing not relief, as she and Percy had anticipated, but a strange prequel to sorrow.

Mary had come to literally despise the place, not only the bay itself, but Casa Magni, where they were living for the summer. Percy had found the house along with their friend Edward Williams and their new acquaintance E. J. Trelawny. They had scoured the region for a peaceful place where they all could live but found nothing worthwhile, other than Casa Magni, which had the look and feel of a fishing shack. Though ramshackle, its one redeeming feature was its position on the hill, which made it possible to watch the sea and to feel as if the waters were leaning into and retreating out of the house.

The back of the house was surrounded by trees so dense that the wind passing through the branches made a sound resembling the crying of children. Combined with the angry waves, these sounds seemed like premonitions to Mary. In this feeling she was like her father, who despised the sea.

The whole place filled Mary with a sense of doom she could not shake off, until she left the house and accompanied Shelley in his small boat. There, sprawled inside the wooden contraption, Mary rested her head in Shelley's lap as he slowly rowed outward, looked up toward a splendidly blue sky, and imagined Casa Magni fading away, along with her fears.

She was not well. Those around her knew it, and she was aware that physically and mentally, she was exhausted. On June 16, she had suffered a miscarriage and the subsequent hemorrhage had left her drained. The ordeal could have killed her, if not for Shelley's swift actions. When he saw the profusion of blood spurting out of her, he lifted her into his arms and plunged her into the icy waters of a bathing tub. Mary must have shivered as the deep chill seeped through her bones and her blood halted its speedy course, all the while seeing the desperation in her husband's eyes and his hurried movements. She had slowly recovered, though she knew it would take longer to heal emotionally. It did not help that Shelley, who seemed tired of her latest bout of melancholy, seemed unable or unwilling to soothe her this time.

Mary also knew she could not rely on her father, who had no patience for feelings of melancholy. A few years earlier, upon the death of her son, William, she had written to William Godwin looking for answers and parental wisdom. Instead, she received a sermon: "I cannot but consider it [Mary's depression] as lowering your character . . . & putting you quite among the commonality & mob of your sex." Unable to offer succor, he became even harsher, plying her with guilt. "What is it you want that you have not? You have the husband of your choice . . . , a man of high intellectual endowments . . . You have lost a child: all the rest of the world, all that is beautiful, all that has a claim upon your kinship, is nothing because a child of three years old is dead! . . . You seem to be shrinking away, & voluntarily enrolling yourself among the worst."

Unlike Mary, Percy adored the area. It was the waters that attracted him most, the boat that he was having built, the simple pleasure of puttering into the sea whenever he wanted to. But lately

Shelley had been dealing with his own issues in the form of odd dreams—or not dreams really, but visions, waking dreams. He was familiar with them because he had suffered from this since childhood. Sometimes the dreams subsided for weeks, but recently they were causing him to scream out in the night and disturb Mary. Not that she had been doing much sleeping herself. Partly due to her physical discomfort and partly due to her spending so much time in bed, her own sleeping pattern had changed dramatically.

She was startled one night when she heard Percy screaming. Fatigued and sore, she got out of bed and began calling his name. But nothing, it seemed, could wake him, until she shook him and he eventually came to and described the dream to her in ways that frightened her as well.

He told her that in his dream their friends Jane and Edward Williams had come to wake him up. When he opened his eyes, he noticed that their bodies had been ripped apart, broken joints poking through their skin, their colorless faces smudged with blood. They were leaning on each other, holding their broken bodies as if for support, and had come to warn him that rushing waters were going to flood the house. In the dream Shelley heeded their advice and ran toward the window, where he noticed a flood of angry water coming toward the house, covering the verandah and eventually engulfing the interior.

Mary shuddered and must have thought about the boat Percy was having constructed, and of their friend E. J. Trelawny, who not long ago had come by the house reciting some of Shakespeare's lyrics he had found so apt for their nautical adventures: "We will all suffer a sea change! We will all suffer a sea change!" Shelley had been charmed by the rhythm of the words and had determined

to have the lines—quoted from "Ariel's Song" in *The Tempest*—
inscribed as a motto on his new nautical toy. But those who knew
the quotation in full, as Mary no doubt did, must have heard in
the words a dire warning:

> *Full fathom five thy father lives;*
> *Of his bones are coral made;*
> *Those are pearls that were his eyes*
> *Nothing of him that doth fade,*
> *But doth suffer a sea-change,*
> *Into something rich and strange.*
> *Sea-nymphs hourly ring his knell:*
> *Hark! Now I hear them—ding-dong bell.*

But Shelley's boat, the *Don Juan*, would never carry the lines
from "Ariel's Song." Rather, the words *Nothing of him that doth fade, but
suffer a sea change* would be etched on Shelley's tombstone in Rome.

Whether E. J. Trelawny knew it or not, whether the Shelleys
even thought of it, he was responsible for their being where they
were and for Shelley's eagerly awaiting the completion of his boat.
This was odd primarily because Trelawny had joined this circle of
acquaintances only months earlier, though he had quickly gained
Shelley's trust.

Until the summer of 1819, Trelawny hadn't even heard of Percy
Shelley and his friends, and then he found himself in the area of
Lake Geneva and was introduced to the name by a local book-
seller, who showed him some of Shelley's works. Not long after,
Trelawny met Thomas Medwin, Percy Shelley's cousin, whose
friendship with Shelley had impressed him deeply. Trelawny then

set out to meet the poet. The opportunity came about sometime later through a friend of both parties, Edward Williams. Williams wrote to Trelawny in April 1821, "Shelley is certainly a man of most astonishing genius in appearance, extraordinarily young, of manners mild and amiable . . . full of life and fun."

In subsequent letters, Trelawny was invited to Pisa, and it was there, in the Williamses' apartment, that he met Shelley. Not surprisingly, Trelawny was stunned by the "mild-looking, beardless boy" and could not help but admire the slightly womanly grace about his persona. Soon Shelley returned and introduced Trelawny to Mary Shelley, who was then twenty-seven years old. Trelawny was struck not so much by her beauty or literary accomplishments, but by "such a rare pedigree of genius." Being the daughter of Mary Wollstonecraft and William Godwin was enough to overwhelm "her own merits as an authoress," but there was also something in Mary's eyes that put him at ease, even though he could not miss the melancholy that seemed a part of her. Trelawny wondered what caused it.

A friendship quickly developed between Trelawny and Shelley, who set out to bring Lord Byron into the mix. Byron was then living across the Ponte Vecchio in Palazzo Lanfranchi. On the day Shelley and Trelawny arrived there, they were received into the palace by the growls of Moretto, Byron's dog. As the two waited for Byron, they walked into a large and opulent room that led directly into a richly decorated parlor room, at the end of which rose a spiraling staircase leading to the upper floors. Trelawny later reported in his memoirs that he noticed right away that Byron limped, though the poet made every effort to hide it by walking quickly toward them.

Unlike John William Polidori, who had tried so hard to be welcomed by the group and was cheerfully mocked, Trelawny was not only befriended while in Pisa but made an essential contributor to their circle. His stories of adventure while in the navy, which for the most part were fabricated, captured the imagination of the poets and their families, who had no problem visualizing the places he spoke of, imagining the people he had met, conjuring the smells he described. In no time, Trelawny was spending his days with the Shelleys and Byron, and his evenings at the Williamses'.

On a daily walk through the harbor, Shelley and Trelawny came upon a collection of boats, and one made in America appealed to them particularly. Trelawny managed to get them onto the boat, and Shelley was so taken with its construction he talked about it and all things related to the sea for days afterward. Trelawny saw such beauty in the boat's lines, he became convinced that only a poet of the sea could have designed it. He proposed that they draw up a model and have boats built for themselves. Shelley, of course, agreed. They drew the lines of their boats in the river sands of the Arno, imagining a colony of English boaters in the Gulf of La Spezia, of which Byron would be a part. Byron also joined them in this dream about the boats and immediately set out to find a large home on the Gulf of La Spezia.

Trelawny hired his friend Captain Daniel Roberts to build the boat. Roberts had risen through the ranks to become a commander in the Royal Navy. His special skills and sociable disposition made him fully capable of building such boats, but he also had the know-how to procure the required permits from corrupt Italian officials. The only thing left was for the Shelleys to find a place to stay.

Byron wanted something similar to his Pisan palazzo, which on the shores of La Spezia was impossible to find, but the Shelleys' needs were far less ostentatious. When they had searched the Ligurian coast, they had only come across Casa Magni. Though the place was deemed unappealing, the Shelleys quickly made arrangements with the owner to move into it for the summer.

For someone who did not know how to swim, Shelley's desire to launch himself into the open waters and to test its mighty powers was foolish. The *Don Juan* was a relatively small boat, barely twenty-five feet in length; Percy certainly should have known it could not withstand the risky currents and strong winds of the Mediterranean waters.

As excited as Shelley was by his new possession, he and Mary still carried a pervasive sense of depression around with them. In June 1822 Shelley sent Trelawny another letter, speaking not only of the boats he and Lord Byron were enjoying, but of a darker sentiment that his mind was ruminating upon: "Should you meet with any scientific person, capable of preparing the *Prussic Acid* or *essential oil of bitter almonds,* I should regard it as a great kindness if you could procure me a small quantity."

Prussic acid was the poison John Polidori had died of. Shelley continued: "It requires the greatest caution in preparation, and ought to be highly concentrated; I would give any price for this medicine. You remember we talked of it the other night, and we both expressed a wish to possess it; my wish was serious, and sprung from the desire of avoiding needless suffering. I need not tell you I have no intention of suicide at present, but I confess it would be a comfort to me to hold in my possession that golden key to the chambers of perpetual rest. The *Prussic Acid* is used in medicine in infinitely minute doses; but that preparation is weak, and has not the concentration necessary to medicine all ills infallibly. A single drop, even less, is a dose, and it acts by paralysis."

Shelley ended his letter by telling Trelawny that Mary was still quite depressed.

An odd dynamic had been set up in the steering of the *Don Juan*. When Trelawny finally visited, he could not help but notice that Edward Williams felt nervous about the boat, making certain everything was in its spot, keeping his eyes constantly on the open waters, so stiff he seemed unable to enjoy the experience. Shelley, on the other hand, seemed completely relaxed, maybe too much so. He steered the boat holding a book in his hands, believing he could read and manage a vessel at the same time, as the activities were "one mental, the other mechanical," therefore requiring different sides of his brain to work.

On July 2 Shelley, Williams, and Trelawny took both boats to Leghorn to care for certain businesses and to buy provisions, having left Casa Magni days before. On the bluff overlooking the Gulf of La Spezia, Mary waited impatiently for her husband's return. Her strange presentiment of doom had not yet left her, so she longed to see the sails unfurl as they docked in the harbor.

The plan was for Shelley and Williams to sail the *Don Juan* across the Gulf of La Spezia, Trelawny following them in Lord Byron's boat, the *Bolivar*. But as they set out to leave, Trelawny was detained in Leghorn for a few days by the harbor patrol due to some irregularities in his paperwork. This did not please Shelley or especially Williams, who was eager to return to Casa Magni and his wife, Jane. After much fanfare, Shelley and Williams decided to set sail on their own. It was already three P.M., an unusual time to depart for such a journey; most boats had left the harbor at dawn. Later Trelawny recalled a mariner standing next to him commenting on the unusual time of departure and giving a dire warning: "They are standing too much on shore . . . The current will set them there . . . she will soon have too much breeze . . . that off-top sail is foolish in a boat with no deck and no sailor on board . . . The devil is brewing mischief."

Trelawny thought it was odd that the old man was talking about a breeze because there was not a hint of it in the air. It was hot and still, and the sailors and mariners were going about their business in a great bustle. "Shifting berths, getting down yards and masts," Trelawny later recalled. "Veering out cables, hauling the ships and quays, boats swelling rapidly to and fro."

Trelawny looked out in the distance at the completely still and calm waters and saw no sign of his friends. But suddenly, the sea turned the color of lead and became as thick as a frozen lake; then raindrops, big hearty ones, started falling from the sky and seemed to bounce off the still surface. The air rippled and thrashed, great noises were heard from the sky, and a terrible gurgle arose from the deepest recesses of the ocean. Trelawny noticed the smaller boats that had not been anchored at the harbor being tossed about like toys while fishermen staggered trying to knot their belongings. It was a fast-moving storm, one of those summer squalls that comes and goes in the span of twenty minutes.

Peace soon returned to the area. Fishing boats and schooners that had been at sea were coming back to their ports. Trelawny hoped to catch a glimpse of the *Don Juan*, thinking that perhaps Shelley and Williams had decided to turn back. But there was no sign of them. He asked those who had returned for information, but no one said they had seen signs of the boat in the open waters.

Trelawny spent the late afternoon and early evening scanning the horizon and the harbor, but as night fell a new storm arose. He returned to the *Bolivar.* The bellowing thunder, the incessant rains, and the constant worry kept him awake, and he resolved that if nothing was heard in the harbor by morning, he would return to Pisa, where Byron was. Perhaps word had come from Casa Magni that Shelley and Williams were safely there. By morning the storm had passed, but all his inquiries about his friends' whereabouts remained unanswered. A terrible feeling of dread had begun to settle upon him, and he hurriedly rode to Pisa, where he found no news from Casa Magni. On Trelawny's arrival, Byron had quickly staggered down his staircase. Trelawny described the events of the past two days, and as he spoke Byron's "lips quivered, and his voice faltered" as he too thought of the possibilities.

At Casa Magni, despair had set in.

Trelawny made his way up the Ligurian coast to Casa Magni. When he got near the city of Via Reggio, he learned that "a punt, a water-keg, and some bottles" had washed ashore and been located by locals. These items were often found on boats, so they did not immediately cause too much concern in him. For several days, parties of searchers tried to find the missing friends, but each day ended with no news. When a week had passed, their worst fears came true. Trelawny heard that two decomposing bodies had been found in the sands near Via Reggio. He rushed to them, wanting to be sure they were the corpses of his two friends.

"The face and hands, and parts of the body not protected by the dress, were fleshless," he later wrote. Right away, he recognized Shelley. It was not so much because the remains were those of a lanky and slight person, but because there was a "volume of Sophocles in one pocket, and Keats's poem in the other, doubled back, as if the reader, in the act of reading, had hastily thrust it away." Edward Williams's body had washed ashore nearly three miles away. His corpse was even more mutilated than Shelley's, the only dignity remaining that of a shirt covering it and the tips of a black silk handkerchief, which Trelawny had seen many times before. Those few bits of fabric are what told Trelawny the body was Williams's. The corpse of the sailor boy who had accompanied them, Charles Vivian, was not found until three weeks later.

Trelawny quickly made his way to Casa Magni, where Mary had been awaiting news of her husband. The two stared at one another. "Is there no hope?" Mary was said to have asked. Trelawny could not bring himself to reply, in essence answering Mary's question.

It was decided that Shelley's remains would be removed from the sands of Via Reggio and interred in the Protestant cemetery in Rome. Williams's remains would be returned to England. But given the length of time that had elapsed and the conditions of the corpses, quarantine laws prohibited the removal of the bodies for fear of the spread of infection. The only way to safely get them out without harming anyone was to cremate them. Oddly enough, Mary did not mind having her husband's body reduced to ashes. Trelawny took it upon himself to arrange the cremations. Williams, it was agreed, would be the first to undergo the procedure.

Trelawny had a funeral pyre built in Leghorn, where the boats had departed. He, Lord Byron, and Leigh Hunt, friend of Byron, were greeted there by a group of soldiers, who were holding iron instruments and shovels, already awaiting him and ready to begin the ordeal. Several feet away a crowd of onlookers had gathered.

The diggers began to shovel away the sand, until "a shapeless mass of bones and flesh" appeared. Byron was seized by the dreadful finality of the scene. "Don't repeat this with me," he was said to have muttered. "Let my carcass rot where it falls."

A funeral pyre made of pines had been set up. Given the dryness of the wood, it caught fire soon after it was lit, the bright heat sending the crowd backward. The flames rose high, then slowly dwindled. As they did, the friends approached Williams's remains and sprinkled them with oil, wine, and frankincense.

Unable to watch this for long, Byron stripped off his clothes and plunged into the cool waters. Trelawny and Hunt followed, and they stayed in until Byron, assaulted by cramps, had to return to shore.

Wooden sticks marked the place where Shelley was temporarily buried. As they approached, they could not help but take note of the desolate beauty of the place, the quiet seashore and endless waters stretching ahead, the pure, blistering whiteness of the sands. It was the kind of solitude Shelley had always desired, the kind he valued. The area possessed such harmony it seemed almost sacrilegious to remove his flesh from beneath the sands, as if they were "vultures." Unlike the previous cremation, when they reached Shelley's corpse no one seemed willing to offer a word, or even capable of doing so.

The diggers began their dreary work until they heard the sound of metal hitting bone and knew they had found Shelley. They removed the body from the sand, the bones not breaking as they had in Williams's case, and placed him on the pyre. They lit the fire and sprinkled wine over it, more "than he had consumed during his life," Trelawny said. As they watched Shelley's body burn, the heat seemed to make their eyes water and waver, the wood snapping and crackling, breaking heartlessly to their ears. Again, the scene was too much for Byron to bear. He returned to his boat. As he swam out, behind him the heat quickly reduced Shelley's body to ashes. But something unusual happened: as Trelawny and Hunt stared into the flames, they noticed that Shelley's heart had not incinerated. Fittingly enough, it remained intact, a complete organ, as if still alive. Trelawny dashed toward the pyre and with his bare hands removed Shelley's heart. His own limbs were singed. In time, Shelley's heart would cause conflict between Edward John Trelawny and Mary Shelley. But for now it was a final remnant of what had been Percy Shelley.

Shelley's ashes were collected and placed in a simple box to be buried in Rome. Not long after they had all left La Spezia and returned to Pisa, the *Don Juan* was found. Captain Daniel Roberts informed Trelawny in a letter dated September 1822: "We have got fast hold of Shelley's boat, and she is now safe at anchor off Via Reggio. Everything is in her, and clearly proves that she was not capsized. I think she must have been swamped by a heavy sea. We have found in her two trunks, that of Williams, containing money and clothes; and of Shelley's, filled with books and clothes."

But as the boat was further inspected, Captain Roberts became suspicious about what had happened. He noticed that many of the timbers were broken, as if another vessel had rammed into it. He shared his views with Trelawny in a letter he dispatched shortly after the first one: "On a close examination of Shelley's boat, we find many of the timbers on the stairboard quarter broken, which makes me think for certain that she must have been run down by some of the feluccas in the squall." A felucca was a large sailing boat constructed of wood that often traveled the waters of the Mediterranean during the summer months. No one could be certain if one had slammed into the *Don Juan,* but Trelawny recalled that his many inquiries had gone unanswered on the day of the storm, and many of the sailors had been reluctant to talk to him about what they had seen, if anything, along their journey. They had adamantly refused to admit they had noticed the boat or Trelawny's friends, which made him think something more might have occurred. Mary Shelley, in time, also came to believe that there had been more to the accident than was discovered.

THE GROUP THAT GATHERED ON THE SHORES OF THE GULF OF LA Spezia disbanded after Shelley's death. Trelawny, who in a certain way had instigated the direction their summer would take, left the area. If he ever thought himself guilty for what had occurred, he never said so. On the contrary, he felt that part of Shelley's demise and unhappiness fell squarely on Mary. Trelawny had come to believe that she had drained him. In a letter he wrote to Claire on December 28, he made his feelings, perhaps unjustly, known, with the same acidity with which William Godwin had shared his: "As to Mary Shelley, you are welcome to her. She was the Poet's wife as bad a one as he c'd have found . . . She was conventional in everything."

Not ready to return to England yet, Mary moved to the outskirts of Genoa, where she rented a house called Casa Negrata. Those who had become close to her also began to disperse. She had come to believe she and Jane Williams might forge a strong friendship to bridge the gap of widowhood. But Jane would have none of that: she returned to England right away. Soon thereafter, Jane began a relationship with Thomas Jefferson Hogg, Percy

Shelley's Oxford classmate and Mary Shelley's ambivalent friend and possible lover during her early years with Percy. By the time of Shelley's death and Jane's involvement with Hogg, Mary's own feelings for Hogg had changed. She no longer felt the kinship that had prompted her to write to him about the death of her first child. Instead, she now truly disliked him, considered him a selfish individual for wanting to capitalize on his friendship with her husband. That Jane became his lover disturbed her, though the affection she felt for the other widow overruled those feelings she had for the man. But that affection didn't last long.

Soon after Jane Williams's affair came to light, Mary also learned that Jane was badmouthing her to her closest friends and acquaintances. Along with Hogg, they were revealing secrets and details of her marriage with Shelley, of the supposed pain Mary had caused her husband, of how depressed he had been prior to his death. Mary became aware of this and felt that an irreparable break in their friendship had occurred. In the years to follow, Jane Williams asked for Mary's forgiveness, which she granted, but the warmth of their earlier relationship was gone.

William Godwin learned of his son-in-law's death not from his daughter but from a family friend, Leigh Hunt. He was irked that Mary had not written to him right away, as he suggested in the letter he quickly dispatched. But perhaps now her suffering, he said, which he learned she was taking "better than could have been imagined," might bridge the gap that had formed between them. Godwin had also been aware of Shelley's odd feelings toward death, or, as he put it, that he had always been "in constant anticipation of the uncertainty of his life, though not in this way."

MARY WAS NOT THE ONLY ONE WHO FELT SHELLEY'S LOSS; LORD Byron did as well. He removed himself to the village of Albaro, a small collection of houses not far from where the Shelleys had lived. Casa Saluzzi was actually a voluptuous palace overlooking the Ligurian coast. As he stood on his balcony, Byron could watch the myriad of changing hues in the waters below him, hear the screeching of the seagulls as they flew above him, and stare at the boats undulating near and far. If he dared to look to his right, he could see the *Don Juan,* anchored to its post. The sight made him nauseous.

He often thought of his friend, and of Mary, of the night of Shelley's disappearance, when she had pounded on his Pisan door, "pale as marble," looking for answers. The fear he had seen in her eyes, the horror she had displayed, was something he had not witnessed before. Nor could he forget it now. He thought about the lost persons, about Mary, and about his own end, which he believed was not that far in coming.

Like Mary's, Lord Byron's natural demeanor tended to darken to melancholy. And in a world that was out of control, he took charge of what he knew best: his regimen of eating. He returned to a strict method of dieting: few vegetables and water, cookies, no meat, and a heavy course of purgatives. In no time he lost the luster he had regained, his clothes began to hang from his bony frame, and the pallor returned to his gaunt features. He also decided he would go to Greece. London had decided to aid the Greeks in their war of independence, and Byron was to become one of their supporters. This new course of action gave him a certain amount of motivation, and preparations were soon made for the voyage.

The only cloud hovering on the horizon was that Byron, like

Mary in the days leading to Shelley's death, began to experience premonitions, which caused him to believe he would not be returning from the islands. To combat that outlook, he engaged Trelawny to assist him, tempting him with the revelation of a new boat he had engaged, the *Hercules*. The name was appropriate given the Greek adventures, but quite ironic considering that Byron neither looked nor felt Herculean.

Trelawny always maintained that "it would have been difficult to find a man more unfit for such an enterprise" than Byron. Aside from having a great name and money to spare, Byron's poor health and lack of stamina made him an unlikely candidate to spend a season in the harsh mountainous regions of Greece.

After traveling through the islands, Byron settled in the small village of Missolonghi. A collection of houses "situated . . . on the verge of the most dismal swamp," it was not a pleasant place to be, and with winter at its doorstep, the dismal rain that drenched the area and the fog that rolled in from the ocean and settled upon the rooftops and steeples only made it worse. Byron quickly fell into a state of despondency and loneliness. He restricted his diet even more, becoming thinner and thinner, waking up only to write letters, quickly losing the enthusiasm he had had for the journey.

Trelawny left after a time, and Byron remained at the mercy of the few servants and several doctors in the area, none of whom seemed to recognize that Byron's eating habits were becoming dangerously unhealthy.

By April, the weather had changed little, but Byron's mood had worsened. Having spent a winter housebound, he yearned to relieve the tension in his limbs and his mind. When the weather cleared, accompanied by his servant Pietro, he rode into the wild country for a few hours of unbridled pleasure. On their return gallop, thunder broke out and the sky opened up, soaking Byron and Pietro. On reaching Missolonghi, Byron began to complain that his body felt odd.

Before long Byron took to his bed with a fever. Soon his body began to tremble and his tongue moved incongruously. He no longer made sense. With doctors by his side, delirium took over, and on April 18, he became unconscious.

The next evening an incredible storm settled over the village. The ebony night came alive with yellow streaks, and booms echoed across the mountains, frightening the young and the old alike. In a home nearby, on hearing a particularly loud crack of thunder, Byron opened his eyes and stared at those surrounding him. For a moment it was believed he had come to, thanks to the noise and harsh pummeling of the rains, his wild fit punctuated by a period of lucidity. But then his lids faltered, he closed his eyes, and he never reawakened.

TRELAWNY LEARNED ABOUT BYRON'S DEATH A FEW DAYS LATER AS HE made his way to Missolonghi. He had been surprised that Byron had chosen to settle in such a horrendous place. This "mud-bank" was desolate, its location too dismal and wet for someone of a delicate disposition. Byron had been lionized during his lifetime, which made it all the more odd that he would die on this disagreeable spot.

He found the house entirely empty and was told that people had been traipsing to and fro in the rooms looking for money Byron might have left. The coffin was in an upstairs room, and

Trelawny later remembered that the poet was "more beautiful in death than in life." He learned that immediately following Byron's death, chaos had ensued as the doctors feared they had been somewhat responsible for his demise. During Byron's delirium, they had subjected him to bloodletting, which was not a smart thing considering that Byron was already weak and malnourished. Bloodletting might have hastened his death. The doctors had also decided to perform a crude autopsy to see if they were to blame. They cut the body to pieces in an attempt to figure out the true cause of death, but in the end, they didn't learn anything.

Trelawny also added to Byron's final indignities by nearly plundering his body. When he was left alone in the room, he removed the sheet covering the corpse and took a peek at his feet and legs, those limbs that had always given the poet such sorrow. "Both his feet were clubbed, and his legs withered to the knees," Trelawny later wrote. "The form and features of an Apollo, with the feet and legs of a sylvan Satyr."

Byron's body was placed in a wooden coffin, which was put into a larger receptacle for its transport back to England. The trip took nearly three months, and when it arrived at the home Byron had left in 1816, the inner casket was taken out and the lid lifted for those few friends and family members who wished to say a final good-bye.

Those who were there, including Mary Shelley, were surprised, because the beauty that had caused such a fuss in life no longer existed. In its place lay a hacked and embalmed body that had been patched together by eager Greek doctors and then further damaged by three months at sea. When the funeral procession made its way by Highgate Hill toward Nottingham, Mary peeked behind the curtains so she could watch its progress. She did not want to attend the funeral, though she did stand by to see it go. As it passed, she realized that Byron's death brought an era to a close. Only she and Claire remained, and she felt her days were also numbered.

Percy's death had brought her grief, but it was combined with a crushing sense of guilt. In the months prior to the accident, Mary had directed a lot of rancor and unspoken hostility at Percy, her anger stemming from her children's deaths, the responsibility for which she directed squarely at Percy. He had suffered greatly from his childrens' deaths, but equally as bad was the distance that had grown between him and Mary. His death evoked in Mary a sense of atonement or punishment. She vowed that from that day forward her job would be to bring all of Percy Shelley's works to publication.

She had wanted to stay in Italy, but finances had prevented that. On returning to England, she discovered something peculiar: her book, *Frankenstein; or, The Modern Prometheus*, had enjoyed a life of its own, had thrived, and as a consequence she was now quite well known. Following *Frankenstein*'s publication, many people tried to figure out who had written the book. A writer from the *British Critic* finally determined that the author was a woman. But it was an anonymous writer in the *Literary Panorama* who first pointed to Mary Shelley: "We have some idea that it is the production of a daughter of a celebrated living novelist," he wrote. The book had also been made into a successful stage play that was running at the English Opera House. She had gone to see it in September 1823 and had been given a warm reception. It must have seemed odd to her, as she sat in the theater and watched her characters—who, she had ada-

mantly convinced herself and others, had come to her fully formed
in a waking dream on the shores of Lake Geneva in 1816—come
to life. She and the others watched as Victor Frankenstein fiddled
with his instruments, as he tried to impart life to his creature, and
then, among the screeching, thrashing, booming sequence of theatrical flashes and pounding, as the creature awakened.

Her goal on returning to England was to use her writing to
achieve financial independence for her and her son, and she set out
to do that. In 1824, she began her second-most acclaimed novel,
The Last Man, which was published in 1826. A deeply depressing
book, the novel takes place at the end of the twenty-first century
during a time when a great and deadly plague has overtaken the
world. Only one man survives, the last man of the human race. In
her journal of May 1824, Mary Shelley wrote, "The Last Man. Yes,
I may well describe that solitary being's feelings, feeling myself as
the last relic of a beloved race, my companions extinct before me."
Mary had no idea what plagues were, where they came from, how
they were spread, what their symptoms were. The book is a work
of the imagination. Still, the deadly consequences depicted in it are
uncomfortably similar to those of real-life plagues studied today.

But another and far stronger theme in *The Last Man* is the role of
the imagination, or rather, the failure and collapse of the imagination. In the book, the imagination works not as a gift, but as a de-
tractor, a corruptor. It was not by coincidence that she intertwined
these two themes: feeling alone and abandoned, the life she had
imagined with Shelley had not come to fruition and now a starkly
different reality had set in.

She also reworked and annotated *Frankenstein* several times. In
1821 she made several corrections to the text. A copy of the book

in which she marked her corrections is housed in the Morgan Library in New York, and Mary's notes and additions are there in the margins in her neat and legible handwriting. But those changes were not incorporated in any future editions. The most famous new edition, and the most controversial one, was published in 1831.

By 1831, her life had changed dramatically. Three of her children had died, and so had her husband, Lord Byron, and Polidori. She was no longer a teenager following a lover across Europe and toting a small child with her, telling stories amidst friends and poets, but a mother to a growing son, a woman in her thirties, a widow, a writer trying to make a living. The idealism of youth had passed and the realities of impending middle age were setting in. She was given the opportunity to revise *Frankenstein* for the New Standard Novels Editions, and to add an introduction. As expected, given that her life had morphed, those of her characters followed suit.

In the 1818 edition, Victor Frankenstein decides on a course of action based on his own whims and desires. Having a mind of his own, he chooses to commit a terrible act against nature and the gods—he learns all he can about alchemical actions; learns all about galvanism; raids cemeteries; builds a creature; abandons it; does not provide for it, either physically or mentally; and in the end he even suggests to Captain Walton that he should take care of the creature after his own death. In essence, all that happened is the direct result of Victor's own actions. But in the 1831 version, Mary Shelley not only mollified her position but actually shifted it: Victor Frankenstein is not responsible for his actions; fate is. His steps were predestined by a creator. Thus, he's no longer a creature with thinking capabilities, but rather, more of a puppet simply performing tasks that have been set out for him. In this way, what he does and the results of those steps become more for-givable. He is not entirely to blame.

Once again, this was not by mere chance. Mary felt that she was no longer a participant in her own life, but rather someone who simply followed a previously written plan. How else could she explain all that had happened? All those who had perished? It had to have been fate.

The introduction also served another purpose. She wrote it, she said, to answer those who often asked her how such a young girl could have come up with this tale and where she'd gotten the idea. In practical details, she explained how the story came to be, writing at length about her so-called waking dream. She was often pestered with those questions from skeptical inquirers who wanted to get to the bottom of how a teenager entangled in a relationship with a married man could have written a book that was becoming a classic.

Thus she told of Villa Diodati, of the ghost story competition, of the conversations about reanimating the dead, of going to bed during a stormy night and coming face-to-face with Victor and his fiend.

There was no one to contradict Mary, no one to say the events she was describing had not taken place or hadn't taken place in the sequence she remembered. Thus, in the 1831 version of *Frankenstein*, she managed to do several things: not only to create the myth of *Frankenstein*—a mad scientist creates a terrifying monster that he unleashes on the world with deadly consequences—but also to give the world another story, the legend of how *Frankenstein* the book was created, that of the author meeting her muse.

EPILOGUE

Though you seek to bury me,
Yet will you continuously resurrect me!
Once I am unbound, I am unbounded!

FRANKENSTEIN UNBOUND

THE EARLY DAYS OF APRIL 2004 WERE GRAY AND somewhat chilly in New York City, giving little indication that spring had begun. Despite the somber atmosphere, a group of family members dressed in warm clothes had gathered early one morning in Central Park, whispering to one another and linking arms together for support. There was nothing peculiar-looking about them; they appeared to be a group of quiet tourists taking in the sights before the whole of the city woke up. They clutched enormous Starbucks coffee cups in their hands, holding them tightly as if they were trying to transfer the warmth into their limbs.

But the paper cups did not hold foamy lattes or the latest overpriced coffee concoction. They were filled with human ashes.

Not long before this morning, the patriarch of the family, Alistair Cooke, had died after a long battle with lung cancer, which had eventually spread to his bones. Having reached the remarkable age of ninety-six, following a famed career as a broadcaster and longtime host of *Masterpiece Theatre*, he left behind specific instructions about how he wanted his body dealt with after his death. He wanted to be cremated and have his ashes scattered along Central Park, a place he had gazed at from his rent-controlled apartment on Fifth Avenue and that he had come to adore in his adoptive home of New York City.

When the end came, on March 30, there was little else for the family to do but to request the assistance of New York Mortuary Services. An employee of the firm appeared hours later to remove the body and took it away for the procedure. The ashes were then returned to the family two days later, and together the relatives began to discuss how to fulfill the request.

Although aware of a local city law prohibiting the scattering of human remains in public places, the family gathered that April morning, coffee cups in hand, and with ceremonious gestures they hoped no one would notice, relegated the ashes to the elements.

The family members then returned to their homes, some in New York and others around the country. Cooke's daughter, Susan, who lived in Vermont, began to find solace in her work as a minister for her community. It must have been comforting for her to think of her father not so much as the frail old man he had become toward the end, but as the spiritual being he had morphed into after his death.

Eighteen months after Cooke's death, the family reunited again, this time in Susan's home in Vermont to celebrate the approaching Christmas season. The region was so still then, and its many lakes were frozen and enchanting. From Susan's home, it was possible to see the distant mountain peaks covered with a snow so fine it seemed nearly transparent. The quiet of this austere landscape had a certain beauty and tranquility that was comforting. That is, until the telephone rang.

As Susan picked up the receiver, she was surprised to hear a woman's voice asking about her father. He had been famous in this country and abroad, and he had a large following that still made its presence known. Perhaps this woman was a fan who had not heard of his passing or a writer working on a piece wishing for an anecdote or a quote. But as it turned out, she was Patricia O'Brien, a detective from the New York Major Case Squad, and she was investigating, among other things, the death of Susan's father. Susan explained that Cooke had died of cancer, but the detective made it clear she was interested not in the way he'd died but in the way his body had been disposed of afterward and how his remains had been handled.

The detective said she had been investigating fraud in a funeral home in Brooklyn and had come across Alistair Cooke's death certificate as well as other documents about his death and the handling of his body after the pickup.

She said the funeral home had not completed the cremation, as the family had requested, but rather, the corpse had been crudely dismembered by a band of so-called body snatchers and cutters. The body parts were later skinned and the remaining bones sold to various companies that cleaned them and processed them, eventually shipping them to different hospitals across the country, where they were implanted into patients undergoing a variety of medical procedures, such as joint replacement, heart valve surgery, skin grafts for burn victims, sports injuries, and even cosmetic procedures. Though Cooke had been old and terminal, new documents had been drawn up for him, with a new age listed, as well as a new cause of death. Having taken what they needed from him, the funeral home had cremated the rest, possibly with the remains of other bodies. The detective could not be sure whether the Cooke family had received their father's ashes or someone else's, or, for that matter, whether they were human ashes at all.

In retrospect, Susan *had* had a funny feeling upon opening those coffee cups so many months ago. As she inserted her fingers into them, the ashes had seemed a little too fine, like talcum powder. She had been told that human remains were grittier, made so by tiny, almost invisible shreds of bone still in them. Susan had reasoned that perhaps New York's mortuaries used bigger, hotter, more modern ovens that completely pulverized a body. She had quickly put it out of her mind.

But as she heard this news about her father, she turned from the sadness she felt upon hearing what his body, albeit dead, had endured to what might have occurred to those who had unwittingly become the recipients of his body parts: Cooke's bones had been too old, too brittle, and most especially, too riddled with disease.

Outside her window, the landscape no longer appeared so benign. If anything, the dark clouds hanging overhead must have seemed ominous, as if they were something dark and corrupted, about ready to burst open.

THOUGH ENGLAND PASSED THE ANATOMY ACT IN 1832, FOLLOWING the Burke and Hare episode as well as the murders committed in London by Bishop and Williams, America's laws on such matters were implemented on a state-by-state basis, beginning with Massachusetts in 1831. The commonwealth's regulations gave medical schools the right to use the bodies of the unclaimed for dissections. These bodies included the ones left in hospitals, workhouses, and tenement houses, and usually belonged to the poor whose families could not afford a burial. Other states quickly followed Massachusetts's lead, in essence blunting the need for body snatchers.

But as the next three or four decades rolled by, new medical schools sprang up across the country, leading to more students wanting to become doctors, thus increasing the demand for more bodies. As such, the body snatchers entered the business again with abandon. When lawmakers became aware of this, they passed new laws designed to prevent a case similar to Burke and Hare's from occurring in America.

By the 1920s, most states had strict laws that regulated the appropriation of corpses for medical use, restricting it to bodies left unclaimed in hospitals, in workhouses, or on the streets. These laws were supposed to allow schools to continue their practices by providing them with enough cadavers; prevent body snatchers from embarking on their ghoulish, if profitable, enterprises; and prevent people from fearing that their bodies would be disinterred and sold to the highest bidder, or, worse, killed and sold to the highest bidder.

Few in recent decades believed that such individuals still existed. But as Alistair Cooke's case showed, body snatchers are still alive and doing well. The band of cutters who defiled Cooke's body—later found, arrested, and convicted—was no better than its counterpart of centuries ago. Granted, their retrieval methods had morphed, their skills had developed and progressed with time, they had learned to blend with the rest of the community, but their goal remained the same: to profit from the dead. They saw nothing unusual in this and saw mortuaries as simply a place where they could cash in.

During her time, Mary Shelley prophesied this despoiling of and profiting from the dead. In *Frankenstein*, Victor Frankenstein says, "A Church yard was to me merely the receptacle of bodies deprived of life." He saw the corpses as a means to an end (to make the creature and become a creator).

That Shelley understood this in the early 1800s was no fluke. The soirées her father, William Godwin, threw every Sunday afternoon provided an opportunity to discuss the latest experiments

involving corpses. Certainly poetry and philosophy were also discussed, but a great deal of time was spent talking about electricity
and galvanization, both of which used corpses for their experiments. Anthony Carlisle, Godwin's friend, witnessed several attempts at the galvanization of bodies, including, it is believed, that
of George Foster by Giovanni Aldini, which Carlisle talked about
often. Humphry Davy, who was a regular at the Godwins' home,
was also a regular user of the latest electrical machines of the era
and became notorious throughout the London community for his
flamboyant experiments.

And the closest one to Mary, Percy Shelley, knew all too well
the effects of voltaic electricity and the experimentations happening on dead bodies; he spoke to her of his days with his cousin
Charles Grove, roaming the halls of the hospital where he worked.
In this way, by being surrounded by these men and hearing their
stories, Mary came to learn the value the dead had to the living.
She was also able to make an educated guess that the demand for
dead bodies would not really wane over time, but would increase.

As Mary Shelley returned to London after her husband's
death in the early 1820s, such ideas were no longer center stage in
her life. Although still interested in the philosophical debates that
had captured Percy Shelley's and her father's imagination, she was
focused on more tangible ordeals. Still guilt-ridden over the suffering Percy had endured just before his death, she was now without
money and without many friends, a fact that pained her. She continued to correspond with Isabella Baxter Booth, the childhood
girlfriend whom she had met while visiting Scotland.

Writing to her, Mary told Isabella about her desperation and disappointments: "When I think of my melancholy return to England . . . ," Mary wrote to Isabella some years later, "of the natural interest one would suppose a young widow with an infant son—the heir of a good fortune might inspire—& the solitude & friendlessness of my position . . . not a human being wd hold out a finger—nay . . . show the inexpensive kindness of an invitation or a kind glance."

Instead of attending lectures on electricity and galvanization, as she had done with Percy and Claire, she threw herself into writing, not only her own original works, but the revisions of the works of her husband, Percy Shelley. She had promised that upon his death, she would bring his poetry to publication, and she achieved that goal in 1839, when the four-volume *Poetical Works of Percy Bysshe Shelley*, the two-volume *Essays, Letters,* and the one-volume *Poetical Works* were published. But while she worked arduously on this task, her health deteriorated. Not infrequently, intense headaches plagued her, and she was often confined to her bed. She attributed these to the mental strains she had subjected herself to and believed they would pass.

With the publication of Percy's works, several things happened: not only was her guilt eased, but her financial situation also improved. With some money at her disposal, she could now set out for her beloved Italy.

AS A YOUNG WOMAN, HER DESIRE HAD BEEN TO GAIN FINANCIAL INdependence and to live in Italy, perhaps permanently. Unfortunately that did not happen for almost twenty years, but now it was possible, in the company of her grown son, Percy Florence. Percy was by then attending Trinity College, which he had entered in the fall of 1837, something that delighted and relieved Mary, as she had come to think that Percy Florence had no aptitude or passion for anything. Those who knew them both, but most especially those who had come to know Percy Florence, had also wondered where his distinctions lay, because it was not clear what his talents were, if he had any.

The barrister Henry Crabb Robinson, who bumped into mother and son in 1839, was struck by the fact that Percy Florence possessed his father's beauty but not much else. Crabb wrote in his diary: "Mrs. Shelley came in, and with her son, a loutish-looking youth, quite unworthy of his external appearance [of] his distinguished literary ancestors. If talent descends, what ought he not be, he who is of the blood of Godwin, Mrs. Wollstonecraft, Shelley, and Mrs. Shelley." But then he added what others had already reported, "Of his moral character is highly spoken of. Of his abilities nothing is said."

Percy Florence also inherited his father's love of the sea and sailing, which did not please Mary. Even so, Percy continued to sail, becoming good enough at it to win several regattas. On an earlier trip to Italy, he had dragged his mother, frightened and whimpering, into a boat and out on a sea undulating with waves.

During the years between 1840 and 1843, Mary, Percy Florence, and several of his school friends finally embarked on their much-talked-about trip to Italy. They tracked through Germany once again, a country she had despised as a youth while traveling with Percy and Claire Clairmont and that she now reviled even more. She nitpicked about the service and the servants, the dingy dining halls, the dirt that greeted them. Her disposition soured even more as they crossed the Brenner Pass; a steady drizzle and dismal fog embraced the town there. It was only the thought of reaching her beloved Italy that kept her going. And on arriving on its soil, she drank deeply from the landscape, which seemed to welcome her as it had once welcomed her and Percy.

As they neared Venice, she could not help but relive her earlier entrance into the city, the one where her daughter, Clara, had died. She recalled the child clutched to her breast, gasping for air in the sweltering carriage as they rushed to meet her husband. But years had now passed, decades, and she decided she should not "dwell on the sad circumstances" of her first views of the city, when "death hovered over the scene."

But despite her best intentions, the events that occurred so long ago were still as fresh as ever, and a mountaintop, a crumpled road, a lone tree, had the power to ignite those inner, fetid passions. Even a shadow, the sun moving across the horizon and peeking through a window, which then cast a light on an object on the table in her home, had the power to remind her of days gone by, of those whom she had loved. In those moments she found that her "nerves . . . [were] strung to their utmost tension by the endurance of pain, or the far severer infliction of mental anguish." But she vowed to think only of the happiness of being in Italy.

She was certainly not surprised to find Venice as unappealing as she had years ago. The buildings, old and crumbling, did not charm her. As she stood by the windows of her room watching twilight fall over the city, the sun setting behind some nameless palazzo whose shadow melted into the murky canals, she had to steal a glance below at the narrow passageways, or farther away, toward the Rialto, and there, in those shadowy places, imagine her husband returning to her after a day spent with Lord Byron. The view had not changed in the least, she reasoned, "but as the Poet says—the difference is to me!"

It was in Rome that the full force of her past, and its losses, came to her. There were so many places where, she said, "the treasures of my youth lie buried." She arrived there during Holy Week, a particularly austere time for Catholic Italy, and the solemnity of the city overwhelmed her. The mournful hymns sung in all the churches were "solemn, pathetic, religious," though they seemed to echo her own deep-seated grief, elevating it to something like spiritual rapture. Sitting in those churches mourning the lost and the losses, she knew Percy Shelley would have understood.

Throughout the trip, Mary was faced with the memories of her lost youth and the lives that had gone with it. She became so overwhelmed she began having relentless headaches. They had started in 1839, and she had believed they were a part of the creative process—she had worked too hard, stressed too much over the publication of the poems; the pressure in her head would abate when the work ended. But as time passed, it became obvious that something more insidious was occurring.

From Rome, she wrote to Claire Clairmont: "It made me suffer in my head, as it were pressure in the brain, more than I had ever done before—& that accompanied by mingled agitation & depression of spirits."

The pains continued when she returned to England: "My brain is so weak, that when a thought touches it, it absorbs it & deprives me of my reason," she wrote again to Claire. Her headaches were having a strange effect. They were not only painful, but they took away the few pleasures she had always counted on in life: reading and writing.

As her physical pains increased, people also began to add to her suffering in ways she had not anticipated. Following Percy Shelley's death, those who had known him tried to capitalize on the poet's notoriety and scandalous lifestyle. One of them, the poet's cousin Thomas Medwin, decided to write Shelley's biography and wanted to include details on Shelley's custody battle with his first wife's family over the future of his children Ianthe and Charles. Medwin asked Mary for her help and attempted to bribe her for particulars, but she refused. "I vindicated the memory of my Shelley and spoke of him as he was," she wrote to Medwin. "An angel among his fellow mortals—lifted far above this world—a celestial spirit and taken away, for we were none of us worthy of him—and his works are an immortal testament giving his name to posterity in a way more worthy of him than my feeble pen is capable of doing."

She didn't want Shelley's name tainted by Medwin, or anyone else. Of course, as the years passed, Mary's memories of her husband had morphed and her guilt had grown, causing her to distort him to such a degree that in her mind he was an angel and not the man who had made her suffer. But she didn't care.

Edward John Trelawny, who had been a friend at Leghorn and taken care of the details when Shelley died, wrote a book titled *Records of Shelley, Byron, and the Author.* He was so enraged by Mary's refusal to help that he described her in the book in ways that were painful to read: "Her capacity can be judged by the novels she wrote after Shelley's death, more than ordinarily commonplace and conventional. Whilst overshadowed by Shelley's greatness her faculties expanded; but when she had lost him they shrunk into their natural littleness. The memory of how often she had irritated and vexed him tormented her after existence, and she endeavored . . . to compensate for the past."

By the mid-1840s, Mary was simply too tired and sore. "I continue very uncomfortable—unable to walk or make any exertion. The doctor declares it to be all Neuralgia," she wrote to Jane Williams Hogg. "I am well in all respects except it is as if the spine were injured the nerves are all alive—it is as if the spine wd altogether give up the ghost."

The one thing that brought her pleasure was Percy Florence's marriage in 1848 to Jane Gibson St. John. Mary was delighted by this union because it provided her with the strong, solid female companionship for which she had yearned her whole life.

By January 1851, her headaches had become uncontrollable and often were accompanied by paralysis of the limbs. By the end of the month, her state became desperate, though doctors always assumed her case was not dire. On February 1, 1851, at twilight, Mary Shelley passed away in her home on Chester Square at the age of fifty-three. The doctors believed she died from a brain tumor.

Percy Florence quickly dispatched letters to family and acquaintances. "About a fortnight ago she had a succession of fits, which ended in a sort of stupor in which she remained for a week—without any sign of life but her breathing which gradually ceased without any pain," he wrote to Mary's friend Isabella Baxter Booth. "And now she has left us most mournful and wretched."

On February 15, *The Athenaeum* ran an article eulogizing Mary Shelley:

> *After having some years disappeared from the world of literary oc-*
> *cupation, the daughter of Godwin and Mary Wollstonecraft, the relict*
> *of the poet "Adonis," died the other day—*
> *Her first work—written during her residence abroad, and the*
> *only one, we believe, referable to the period of her married life—was*
> *"Frankenstein" which scared and startled the world by its preternatu-*

*ral power, promising further inspiration of a wild originality un-
known in English fiction . . . That Mrs. Shelley would never equal her
first effort in poetical fiction, might have been foreseen at the moment of
the tragedy of her husband's frightful death—one of those visitations
the traces of which one never to be affected, and which bereaved the sur-
vivor of guidance, companionship, incitement to emulate for ever—
All of Mrs. Shelley's writings have a singular elegance of tone;—but
all of them a pervading melancholy. Her tales of the world we live in
are unreal in the excess of their sadness;—while in her more romantic
creations . . . with all their beauty, there is blended a certain languor
which becomes oppressive. Hence, most of her works of imagination are
unfairly neglected,—whether, however, such neglect shall be reversed
on a future day or not, her "Frankenstein" will always keep for her a
particular place among the gifted women of England.*

Copyright laws allowed other writers to use previously published works and to interpret them as they saw fit, and even before
Mary died, Victor Frankenstein and his creature had successfully
gone through several stage productions. Given the popularity
of the plays in Europe and the notoriety of the 1831 edition of
the book, it did not take long for the story to cross the Atlantic
and make its way to America. Though many Americans initially
became acquainted with the characters through the book, it was
the 1931 release of Universal Studios' *Frankenstein* that truly galvanized its place in history and in the viewers' minds. Directed by
James Whale, it turned a relatively unknown, middle-aged actor,
Boris Karloff, into a movie star. But the movie did something else:
it stripped down the book to its bare bones and turned Mary Shelley's moral tale into a basic horror story.

In the book, the creature is a conflicted, eloquent, sad entity searching for knowledge, acceptance, humanity, love, and a mate; basically he experiences all the human emotions. Nearly a third of the book is given over to him. We hear his thoughts, feel his pain and anguish; we sympathize with him even though we do not condone the actions he takes as the book rolls along. None of these fine nuances exist in the movie. The creature is simply an oversized and dimwitted batch of bones and skin, pale and long-limbed, grunting his way across villages intent on wreaking havoc on everything and everyone he runs into. His most endearing trait, language, is taken away from him.

From the beginning, viewers believe he can only cause harm. Readers of the book are aware that Victor had found his materials—bones, skin, and brain—in death houses and cemeteries; no one knew to whom those things had belonged. Victor could have taken them from a criminal or a priest, a man or a woman, someone rich or poor, a good person or bad. The creature begins with a clean slate, more or less, and it is up to his creator, Victor Frankenstein, and the world at large to make sure he is raised as a moral and decent individual. He becomes what and who he is through the knowledge and experience he gains from Dr. Frankenstein and the world that surrounds him.

The movie includes no such considerations. The creature is fitted with a criminal's brain (the "normal" brain was dropped on the floor by mistake), which means the creature is marked for criminal behavior from the start—in a sense, the movie's message being that predisposition trumps free will.

But viewers didn't go to see *Frankenstein* to embark on any deep philosophical debates. It was released during the Depression, so people lined up simply for some good old-fashioned entertainment, and the movie provided that. *Frankenstein* turned out to be one of the earliest blockbusters of the cinema era, and as expected, different takes on the tale soon appeared. *Bride of Frankenstein* (1935), *The Ghost of Frankenstein* (1942), *Frankenstein Meets the Wolf-Man* (1943), and *Abbott and Costello Meet Frankenstein* (1948) are just some of the few interpretations that followed the 1931 release, the creature by now a monstrous fiend who invaded nightmares with guttural sounds and long outstretched arms.

It took comedian Mel Brooks to add some softness to the creature in 1974, when *Young Frankenstein*, a blend of music and comedy, was released. Since then, the story has remained popular not only in the movies, but in books for children and adults; comic books; music; Halloween costumes, of course; and even cereals.

While Mary Shelley did not anticipate the impact her story would have on society, *Frankenstein*, and most especially its creature, bridges the gap between the early doings of the body snatchers and the experiments in galvanism, and today's attempts to meddle with nature, also serving as a warning. Are scientists ready for what they might create or uncover?

ACKNOWLEDGMENTS

From the very beginning, everyone I talked to seemed amused and intrigued by Mary Shelley's writing of *Frankenstein; or, The Modern Prometheus*, and the darker story of galvanism and the resurrection men that lurked behind the real one. Many people I contacted—researchers, scientists, writers, and archivists—were more than happy to lend their time and knowledge, and many of them had not only heard of, but in some way or another delved into the topics themselves.

A generous debt of gratitude for conversations on Mary Shelley, galvanism, body snatching, and the 1800s goes to Elizabeth Rettleback, Peter McGowren, Roger Hall, Erik Midelfort, Radu Florescu, Helen Macdonald, Malcolm Archibald, Giuliano Pancaldi, William Eamon, Seamus Perry, Miranda Seymour, Vita Fortunati, Graham Allen, Carlo Ginzburg, Adriano Prosperi, Raffaela Simili, Graeme Gooday, Anne M. Stiles, Marco Bresadola, Andre Parent, Chris Smith, Stanley Finger, Sheldon Glashow, Sherry Ginn, Fabio Bevilaqua, Paula Feldman, Stefani Englestein, Iwan Rhys Morus, Elizabeth Ihrig, Allison Faichney, Paula Bertucci, Anne K. Mellor, Paolo Mazzarello, Charles Robinson, Tom Hoobler, Jan Andrew Henderson, and Amak Mayer.

Many interesting and somewhat disturbing conversations were had on the current trade of cadavers, and for those times and their willingness to share I wish to thank Wendy Kogut, Tiffany Milius, Patricia McNeill, Gil Hedley, Susan Cooke Kittredge, John Gentile, Mario Gallucci, and Todd Olson.

Particular thanks to Kathy Flynn, at the Phillips Library in Salem, Massachusetts; the staff at the Morgan Library and Museum; the Bodleian Library; Louise King, archivist at the Royal College of Surgeons of England; Richard Keenan and Julianne Simpson at the Wellcome Library; the staff at the Archivio di Stato di Bologna; Giacomo Nerozzi at the Biblioteca Comunale Dell'Archiginnasio; Lucia Maranni at the Comune di Bologna Archivio Storico; Antonio Campigotto at the Museo del Patrimonio Industriale; the University of Bologna; Maryon Liscinky at the FDA Office of Public Affairs; Natalie Rosset, research assistant at the University of Dundee; David McClay at the John Murray Collection, and the Trustees of the National Library of Scotland; Sue Hodson, curator of Literary Manuscripts, the Huntington Library; Bruce Barker-Benfield, senior assistant librarian, Bodleian Library; Oxford University Press; and Harvard University Press.

Jake Bauman read the proposal and became interested in it right away, working with diligence and passion to bring it to the eye of the right publisher; for his humor, dedication, and friendship I will always be grateful. And to Rob Weisbach, for taking over where Jake left off and with gusto spiriting it forward.

I am grateful to have found in Henry Ferris at William Morrow a wonderful, compassionate, and passionate editor who saw the merits of the story right away and pointed to its flaws with kindness and generosity, devoting more time to me than I could have wished for or dreamed of. I am deeply grateful to him for the opportunity, and to everyone at William Morrow for making me feel at home.

My thanks to everyone at Emerson College, especially Richard Zauft, who hired me to teach a particular class where *Frankenstein* took center stage in the Institute for Liberal Arts and Interdisciplinary Studies; to Amy Ansell for keeping me on; to everyone in the department for listening to too much talk about body snatchers and such; and to all of my students, past and present, who have been inspiring, fun, and always a source of creativity.

And to my family, here and abroad, for believing this could be done.

NOTES

PROLOGUE

The details of Luigi Galvani's experiments, his dissections of frogs, and the help his family and assistants offered are to be found in several places, including Galvani's *De viribus electricitatus in motu musculari commentarius*. Also, portions of those experiments are reprinted in *The International Workshops Proceedings* from the University of Bologna and the book published by the Cooperativa Libraria Universitaria Editrice Bologna, *Fra biologia e medicina*.

The reference to the "martyrs of science" comes originally from a quote by Hermann Helmholtz, which is reprinted in Frederick Holmes's "The Old Martyr of Science: The Frog in Experimental Physiology."

The details of Giovanni Aldini's experiments on George Foster are to be found in Aldini's *An Account of the Late Improvements in Galvanism: With a Series of Curious and Interesting Experiments Performed Before the Commissioners of the French National Institute and Reprinted Lately in the Anatomical Theatres of London: To Which Is Added, an Appendix, Containing the Author's Experiments on the Body of a Malefactor Executed at New Gate*. Also, references to Aldini's experiments can be noted to some extent in Radu Florescu's *In Search of Frankenstein*, Anne K. Mellor's *Mary Shelley: Her Life, Her Fiction, Her Monsters*, and the *Newgate Calendar*.

Sir Humphry Davy's references can be found in further detail in Jane Fuller's *Young Humphry Davy: The Making of an Experimental Chemist*, as well as Humphry Davy's *A Discourse, Introductory to a Course of Lectures on Chemistry*, and Davy's nine-volume *Collected Works*.

CHAPTER I

The story of Mary Shelley and Claire Clairmont's hiding beneath the family's sofa can be found in numerous places: Betty T. Bennett's *The Letters of Mary Wollstonecraft Shelley*; Paula R. Feldman and Diana Scott-Kilvert's *The Journals of Mary Shelley*; Radu Florescu's *In Search of Frankenstein*; and Anne K. Mellor's *Mary Shelley: Her Life, Her Fiction, Her Monsters*.

The section on Mary Shelley's birth, Mary Wollstonecraft's death, and William Godwin's reaction to the events comes primarily from William Godwin's *Memoirs of the Author of a Vindication of the Rights of Woman*, though references to the events can be found in many other sources, such as William Godwin's "Correspondence & Papers," including his diary, now held at the Abinger Collection in the Bodleian Library; passages in Charles Robinson's *The Frankenstein Notebooks*; Anne K. Mellor's *Mary Shelley: Her Life, Her Fiction, Her Monsters*; footnotes and introductions in Paula R. Feldman and Diana Scott-Kilvert's *The Journals of Mary Shelley*; and Betty T. Bennett's *The Letters of Mary Wollstonecraft Shelley*.

The information on electricity and the thunderstorms on the night of Mary Shelley's birth comes from Humphry Davy's *A Discourse, Introductory to a Course of Lectures on Chemistry*.

The section on William Godwin and Mary Wollstonecraft's friendship and courtship can be found in William Godwin's *Memoirs of the Author of a Vindication of the Rights of Woman*.

References to the reviews of William Godwin's *Memoirs of the Author of a Vindication of the Rights of Woman* come from London's *Monthly Review*.

The introduction to Mrs. Jane Clairmont and the courtship that ensued with William Godwin come from William Godwin's *Diary* and "Correspondence & Papers."

Quotations from Charles Phillips on public hangings come from his own book *Vacation Thoughts on Capital Punishments*, while details on the execution of John Holloway and Owen Haggerty come from G. T. Crook and John L. Ragner's *The Complete Newgate Calendar*, including the announcement and prayer from the bellman at St. Sepulchre.

Humphry Davy's quotations can be found in his *A Discourse, Introductory to a Course of Lectures on Chemistry*, as well as his *Collected Works*.

CHAPTER 2

References to Luigi Galvani's demonstrations, the anatomical theater sections, and the criminality in Bologna can be found in several places: Marco Bresadola's "Medicine and Science in the Life of Luigi Galvani" and "Animal Electricity at the End of the Eighteenth Century"; Giovanna Ferrari's "Public Anatomy Lessons and the Carnival: The Anatomy Theater of Bologna, Past and Present"; Lodovico Frati's *Il settecento a bologna*; Mario Fanti's *Presentazione*; Luigi Galvani's *De viribus electricitatus in motu musculari commentarius*; Marco Bresadola and Giuliano Pancaldi's *Luigi Galvani International Workshops Proceedings*; Robert M. Green's "A Translation of Luigi Galvani's *De Viribus Electricitatus in Motu Muscularis Commentarius. Commentary on the Effect of Electricity on Muscular Motion*"; J. L. Heilbron's "The Contributions of Bologna to Galvanism"; Pasquale Orlandi's *Memorie storiche della terra di medicina*; and Marcello Pera's *The Ambiguous Frog*.

Berengario da Carpi's experiments and demonstrations can also be found in the above sources, including Giovanna Ferrari's "Public Anatomy Lessons and the Carnival."

Andreas Vesalius's visit to and lectures in Bologna, as well as his studies of anatomy and evening escapades into body snatching, can be found in various sources, including C. D. O'Malley's *Andreas Vesalius of Brussels* and Vesalius's *De humani corporis fabris libri septem*, translated by William Frank Richardson.

Baldasar Heseler's recollections of Vesalius's anatomy lessons in Bologna can also be found in the translation of *De humani corporis fabris*, as well as Ruben Erikkson's *Andreas Vesalius' First Public Anatomy at Bologna 1540. An Eyewitness Report.*

Comments and quotations on the early anatomy theaters come from William Brockbank's "Old Anatomical Theatres and What Took Place Therein."

Luigi Galvani's experiments and notes on his demonstrations can be found in his *De viribus electricitatus in motu musculari commentarius.*

Bassiano Carminati's interchange with Galvani and the various letters between the two are also reprinted in Galvani's *De viribus.*

Details on the life of Alessandro Volta, his experiments, and his exchanges with Luigi Galvani can be found in various places: Marco Bresadola's *Animal Electricity at the End of the Eighteenth Century: The Many Facets of a Great Scientific Controversy;* Giovanna Ferrari's "Public Anatomy Lessons and the Carnival: The Anatomy Theatre of Bologna"; Luigi Galvani's *De viribus electricitatus in motu musculari commentarius;* the University of Bologna's *International Workshop Proceedings;* Raffaele Bernabeo's *Luigi Galvani (1798–1998) fra biologia e medicina: Atti della Accademia delle Scienze dell'Istituto di Bologna, classe di scienze fisiche, anno 286;* Orlando Pasquale's *Memorie storiche della terra di medicina;* Alessandro Volta's *Elettricità scritti scelti;* and Marcello Pera's *The Ambiguous Frog.*

Giovanni Aldini's experiments on oxen, his procurement of cadavers, and his meddling with the lunatic Luigi Lanzarini can be found in greater detail in John Aldini's *An Account of the Late Improvements in Galvanism; with a Series of Curious and Interesting Experiments Performed Before the Commissioners of the French National Institute and Repeated Lately in the Anatomical Theatres of London; to Which Is Added, an Appendix, Containing the Author's Experiments on the Body of a Malefactor Executed at New Gate.*

CHAPTER 3

Some descriptions of the city of London upon Giovanni Aldini's arrival can be found in Roy Porter's *English Society in the Eighteenth Century.*

Further information on James Graham's experiments is found in Peter Otto's "The Regeneration of the Body: Sex, Religion and the Sublime in James Graham's Temple of Health and Hymen."

Horace Walpole's reaction to Graham's Temple of Health is printed in Walpole's

"To Lady Ossory," reprinted in *The Yale Editions of Horace Walpole's Correspondence.*

The tales of the resurrectionists can be found in even more detailed descriptions in Christian Baronet's *The Autobiography of Sir Christian, Baronet;* James Moores Ball's *The Body Snatchers: Doctors, Grave Robbers and the Law;* James Blake Bailey's *The Diary of a Resurrectionist, 1811–1812, to Which Are Added an Account of the Resurrection Men in London and a Short History of the Passage of the Anatomy Act;* and Ruth Richardson's *Death, Dissection, and the Destitute.*

The doctor's poem quoted at his deathbed can also be found in James Blake Bailey's *The Diary of a Resurrectionist.*

More extensive information on George Foster's alleged crime and subsequent trial and confession can be found in G. T. Crook and John L. Ragner's *The Complete Newgate Calendar.*

Charles Dickens's visit to the prison can be read in full in "A Visit to Newgate," originally printed in *Sketches by Boz.*

Further and more complete information on the Murder Act can be found in James Moores Ball's *The Body Snatchers: Doctors, Grave Robbers and the Law,* as well as James Blake Bailey's *The Diary of a Resurrectionist.*

George Foster's death by hanging and his subsequent retrieval from the gallows by Mr. Pass are detailed in G. T. Crook and John L. Ragner's *The Complete Newgate Calendar,* while Giovanni Aldini's extensive experiments on the body of George Foster can be found in full in John Aldini's *An Account of the Late Improvements in Galvanism; with a Series of Curious and Interesting Experiments Performed Before the Commissioners of the French National Institute and Repeated Lately in the Anatomical Theatres of London; to Which Is Added, an Appendix, Containing the Author's Experiments on the Body of a Malefactor Executed at New Gate.*

The death of Mr. Pass was reported in Crook and Ragner's *The Complete Newgate Calendar,* as well as John Aldini's *An Account of the Late Improvements in Galvanism.*

CHAPTER 4

The information on Paracelsus can be found in his own book *De rerum natura (Concerning the Nature of Things),* as translated by A. E. Waite.

Portions of Mary Shelley's *Frankenstein* can be found in James Rieger's reprinted edition of *Frankenstein; or, The Modern Prometheus,* as well as the edition of *Frankenstein; or, The Modern Prometheus* with introduction and notes by Maurice Hinoce.

Paracelsus's recipe for the homunculus and accounts of many of his other experiments can be read in full in his *De rerum natura (Concerning the Nature of Things).*

The quote and recipe for making the golem in the Polish Jewish tradition comes from Jacob Grimm's *A Journal for Hermits.*

Details on Percy Shelley's life at Oxford can be found in even further detail in Thomas Jefferson Hogg's *Shelley at Oxford* as well as Edward Dowden's *The Life of Percy Shelley,* in two volumes.

Tiberius Cavallo's experiments and his interchanges with Dr. James Lind are printed in Cavallo's *A Complete Treatise in Electricity, in Theory and Practice, with Original Experiments, Containing the Practice of Medical Electricity, Besides Other Additions and Alterations.*

Percy Shelley's letter to Thomas Jefferson Hogg was reprinted in Edward Dowden's *The Life of Percy Shelley,* as well as T. J. Hogg's *The Life of Percy Shelley.*

A description of Percy Shelley's dabbling with laudanum and subsequent nightmares can be found in Edward Dowden's *The Life of Percy Shelley* as well as Thomas Medwin's *The Life of Percy Bysshe Shelley.*

Descriptions of Dundee's history, including the history of whaling, are to be found in the *Dundee Advertiser,* as well as from the Dundee Whaling History Project.

Physical descriptions of Mary Shelley are quoted from Edward Dowden's *The Life of Percy Shelley,* while Percy Shelley's marriage to Harriet Westbrook, their escape to Scotland, and his subsequent meeting with Mary Shelley can be found in various places, including Thomas Jefferson Hogg's *Shelley at Oxford;* Edward Dowden's *The Life of Percy Shelley;* Thomas Medwin's *The Life of Percy Bysshe Shelley;* and to some extent, Betty T. Bennett's *The Letters of Mary Wollstonecraft Shelley.*

Percy Shelley's letter to T. J. Hogg detailing Eliza Westbrook's involvement in his life and that of his wife can be found reprinted in Edward Dowden's *The Life of Percy Shelley.*

Descriptions of Harriet Shelley after Percy Shelley's abandonment can be read in full in Mark Twain's *In Defense of Harriet Shelley and Other Essays.*

CHAPTER 5

William Godwin's letter to John Taylor is reprinted in Paula Feldman and Diana Scott-Kilvert's *The Journals of Mary Shelley.*

Claire Clairmont's quote on leaving Mary Godwin and Percy Shelley alone in the grounds of the cemetery can be found in R. Glynn Grylls's *Claire Clairmont, Mother of Byron's Allegra* and Marion Kingston's *The Journals of Claire Clairmont* and *The Clairmont Correspondence: Letters of Claire Clairmont, Charles Clairmont, and Fanny Imlay Godwin.*

Details on Claire Clairmont, and Mary Godwin and Percy Shelley's elopement to the mainland, including their trip to France and subsequent return to England, can be read in full in Mary Shelley's *A Six Weeks' Tour Through a Part of France, Switzerland, Germany, and Holland, with Letters Descriptive of a Sail Round the Lake of Geneva, and of the Glaciers of Chamouni.*

Percy Shelley and Mary Godwin's three-hour walk in the German town of Nieder-Beerbach and their likely hearing of Burg Frankenstein and its inhabitants was also detailed in Radu Florescu's *In Search of Frankenstein* and lightly touched upon by Anne K. Mellor in *Mary Shelley: Her Life, Her Fiction, Her Monsters.*

Johann Konrad Dippel's life and experiments appeared also in E. E. Aynsley and W. A. Campbell's "Johann Konrad Dippel," while the experiments he might have read of in order to concoct his recipe for Dippel's Oil can be read in full in Raymond Lully's *Experimenta* and Robert Boyle's *The Sceptical Chymist*.

CHAPTER 6

Details of Mount Tambora's explosion, its effect on the villagers, and the subsequent help that was extended to them can be found in Thomas Standford Raffles's *Memoir on the Life and Public Services of Sir Thomas Stanford Raffles, F.R.S. & C., Particularly in the Government of Java 1811–1816, and of Bancoolen, and Its Dependencies 1817–1824, with Details of the Commerce and Resources of the Eastern Archipelago, and Selections from His Correspondence.*

Percy Shelley and Mary Godwin's new reality in London, including Claire Clairmont's stay in their house, can be found in Edward Dowden's *The Life of Percy Shelley*.

Mary Shelley's journal entry upon the death of her child can be found in Paula R. Feldman and Diana Scott-Kilvert's *The Journals of Mary Shelley*. Reprints of the *Times* advertisement for the brothers Garnerin can also be found in *The Journals of Mary Shelley*.

Mary Shelley's letters to T. J. Hogg are reprinted in Betty T. Bennett's *The Letters of Mary Wollstonecraft Shelley*, where Mary's relationship with Hogg is also analyzed by Ms. Bennett.

Claire Clairmont's comments to Edward Augustus Silsbee on the nature of Mary's relationship with T. J. Hogg can be found in the Silsbee Papers.

Thomas Moore's recollections of Lord Byron are quoted in Teresa Guiccioli's *My Recollections of Lord Byron*, while Lord Byron's quotes and descriptions come from Countess Blessington's *A Journal of Conversations with Lord Byron, with a Sketch of the Life of the Author*.

Lady Byron's recollections and letters about her husband come from a letter she wrote and that was reprinted in *The Passages from Lady Anne Barnard's Private Family Memoirs*.

Lord Byron's comments on his marriage and the birth of his daughter come from Countess Blessington's *A Journal of Conversations with Lord Byron, with a Sketch of the Life of the Author*.

Claire Clairmont's letters to Lord Byron are reprinted in Marion Kingston's *The Clairmont Correspondence: Letters of Claire Clairmont, Charles Clairmont, and Fanny Imlay Godwin*.

John William Polidori's description of his travels with Lord Byron, his letters to his sister, and the subsequent meeting with Percy Shelley, Mary Godwin, and Claire Clairmont can be found in a more extensive form in D. L. MacDonald's *Poor Polidori: A Critical Biography* and William Michael Rossetti's *The Diary of Dr. John William Polidori*.

Fanny Imlay's letters to Mary Godwin can be found in their entirety in Marion

Kingston's *The Clairmont Correspondence: Letters of Claire Clairmont, Charles Clairmont, and Fanny Imlay Godwin.*

Mary Godwin's letters to her sister Fanny are reprinted in Mary Shelley's *History of a Six Weeks' Tour Through a Part of France, Switzerland, Germany, and Holland, with Letters Descriptive of a Sail Around the Lake of Geneva, and the Glaciers of Chamouni.*

Percy Shelley's letter to Teresa Guiccioli is reprinted in Teresa Guiccioli's *My Recollections of Lord Byron.*

Mary Shelley's comments on the thunderstorms visiting the lake can be found in her letters to Fanny Imlay, reprinted in Mary Shelley's *History of a Six Weeks' Tour.*

Mary Shelley's recollections of the evening conversations at Villa Diodati were introduced in her 1831 edition of *Frankenstein; or, The Modern Prometheus.*

John William Polidori's details on his studies and subsequent knowledge of waking dreams and body snatching comes from David Petrain's "An English Translation of John William Polidori's (1815) Medical Dissertation on *Oneirodynia* (Somnambulism)."

John William Polidori's quotes come from William Michael Rossetti's *The Diary of Dr. John William Polidori,* while Mary Shelley's comments on Polidori's idea of a "skull-headed lady" come from her introduction to the 1831 edition of *Frankenstein.*

The description of the evening entertainment at Villa Diodati, including the reading of phantasmagoria and Byron's idea to write their own ghost stories, comes from Mary Shelley's 1831 introduction to *Frankenstein; or, The Modern Prometheus.*

In William Michael Rossetti's *The Diary of Dr. John William Polidori,* the entry of June 18 tells of the reading of "Christabel."

The story of the arrival of Victor Frankenstein and his fiend was first introduced by Mary Shelley in 1831 in her introduction to *Frankenstein; or, The Modern Prometheus.*

CHAPTER 7

Details of the days following the ghost story competition can be found in William Michael Rossetti's *The Diary of Dr. John William Polidori.*

Lord Byron's letter to Douglas Kinnaird detailing his initial involvement with Claire Clairmont was originally printed in Leslie Marchard's *Byron: A Biography,* later reprinted in Paula R. Feldman and Diana Scott-Kilvert's *The Journals of Mary Shelley.*

John William Polidori's letters to his father, Gaetano Polidori, and details of his trip to Italy and the bad luck he found there can be found in William Michael Rossetti's *The Diary of Dr. John William Polidori,* and to some extent, D. L. MacDonald's *Poor Polidori: A Critical Biography.*

John Hobhouse's quote about his involvement in helping Polidori leave the premises is found in D. L. MacDonald's *Poor Polidori: A Critical Biography.*

Accounts of John William Polidori's return to England, his failed attempts at a law career, his enjoyment of bordellos and gambling, and his accident and "brain damage" can be found in several places, such as D. L. MacDonald's *Poor Polidori: A Critical Biography*; William Michael Rossetti's *The Diary of Dr. John William Polidori*; the *Norfolk Chronicle and Norwich Gazette* of September 1817; and the *New Monthly Magazine.*

Further details of John William Polidori's tale "The Vampyre" and the subsequent fuss that ensued over its authorship can be found in D. L. MacDonald's *Poor Polidori: A Critical Biography*; William Michael Rossetti's *The Diary of Dr. John William Polidori*; a letter from John William Polidori to the *New Monthly Magazine*; and Lord Byron's response to the controversy in the *New Galignany Magazine.*

John William Polidori's death was covered in *The Traveller*, August 1817; Polidori's quote about being overshadowed by Lord Byron can be found in William Michael Rossetti's *The Diary of Dr. John William Polidori*, in the entry dated May 28. In Rossetti's book there is also Lord Byron's reaction to the death of Polidori.

Claire Clairmont's letters to Lord Byron, detailing her pregnancy, hopes, and new habitation, are reprinted in Marion Kingston's *The Clairmont Correspondence: Letters of Claire Clairmont, Charles Clairmont, and Fanny Imlay Godwin.*

Fanny Imlay Godwin's trip to Bristol and her subsequent suicide by laudanum are covered in Paula R. Feldman and Diana Scott-Kilvert's *The Journals of Mary Shelley*; Marion Kingston's *The Clairmont Correspondence: Letters of Claire Clairmont, Charles Clairmont, and Fanny Imlay Godwin*; and *The Cambrian.*

People's views of Fanny Godwin are detailed, in part, in Paula R. Feldman and Diana Scott-Kilvert's *The Journals of Mary Shelley*; Marion Kingston's *The Clairmont Correspondence: Letters of Claire Clairmont, Charles Clairmont, and Fanny Imlay Godwin*; Edward Dowden's *The Life of Percy Shelley*; Frederick Jones's *Maria Gisborne and Edward E. Williams, Shelley's Friends: The Journals and Letters*; and E. V. Lucas's *The Letters of Charles Lamb to Which Are Added Those of His Sister, Mary Lamb.*

Details of Mary Godwin's marriage to Percy Shelley come from Claire Clairmont, who spoke to Captain Edward Augustus Silsbee, details of which can be found in the Silsbee Papers.

Mary Shelley's letter to Lord Byron is reprinted in Betty T. Bennett's *Selected Letters of Mary Wollstonecraft Shelley*, while Mary Shelley's entry upon finishing *Frankenstein* can be found in Paula R. Feldman and Diana Scott Kilvert's *The Journals of Mary Shelley.*

Reviews of *Frankenstein* appeared in *The Quarterly Review* of March 1818.

Percy Shelley's "Alastor" was originally published in 1816, then reprinted by Raymond D. Havens in *PMLA.*

Claire Clairmont's letter to Lord Byron hailing the publication of *Frankenstein* can be found in Marion Kingston's *The Clairmont Correspondence: Letters of Claire Clairmont, Charles Clairmont, and Fanny Imlay Godwin.*

CHAPTER 8

The murder of Alexander Love and attempted murder of his grandson by Matthew Clydesdale, Clydesdale's arrest, his subsequent trial, and his attempted suicide, as well as his dissection, are covered in Peter Mackenzie's "The Case of Matthew Clydesdale" and F. L. M. Pattison's *The Clydesdale Experiments, an Early Attempt at Resuscitation* and *Granville Sharp Pattison: Anatomist and Antagonist*.

Andrew Ure's life and public spat with his wife is also detailed in F. L. M. Pattison's *Granville Sharp Pattison: Anatomist and Antagonist*.

Details of Matthew Clydesdale's supposed resurrection are to be found in part in Peter Mackenzie's "The Case of Matthew Clydesdale" and Andrew Ure's "An Account of Some Experiments Made on the Body of a Criminal Immediately After Execution, with Physiological and Practical Observations."

Reports on William Burke and William Hare's vicious crimes, as well as Robert Knox and his involvement with Burke and Hare, appeared in many publications of the time. Further information can be found in *Blackwood Magazine* issues from 1829; Thomas Ireland Junior's *The West Port Murders: An Authentic Account of the Atrocious Murders Committed by Burke and His Associates, Containing a Full Account of All the Extraordinary Circumstances Connected with Them*; James Moores Ball's *The Body Snatchers: Doctors, Grave Robbers and the Law*; G. T. Cook and John L. Rayber's *The Complete Newgate Calendar*; and various editions of *The Scotsman*.

Some information on the Anatomy Act can be found in James Moores Ball's *The Body Snatchers: Doctors, Grave Robbers and the Law*; James Blake Bailey's *The Diary of a Resurrectionist. 1811–1812, to Which Was Added an Account of the Resurrection Men in London and a Short History of the Anatomy Act*; and Ruth Richardson's *Death, Dissection and the Destitute*.

CHAPTER 9

The description of the city of Naples and Robinson's quotes about the city can be found in Henry Crabb Robinson's *Diary, Reminiscences, and Correspondences of Henry Robinson Crabb*.

The story of the Neapolitan baby can also be found in several publications in greater detail, such as Betty T. Bennett's *The Letters of Mary Wollstonecraft Shelley*; Anne K. Mellor's *Mary Shelley: Her Life, Her Fiction, Her Monsters*; and Paula K. Feldman and Diana Scott-Kilvert's *The Journals of Mary Shelley*.

Edward Augustus Silsbee's travels to Italy and details of his conversations with Claire Clairmont are written in the Silsbee Papers, Memo Books, 1875–1877.

Percy Shelley's dream of the house being flooded can be found in a letter from Mary Shelley to Maria Gisborne, dated August 15, 1822, and reprinted in Betty T. Bennett's *Selected Letters of Mary Wollstonecraft Shelley*.

Mary Shelley's recollections of E. J. Trelawny entering her house singing, "We will all suffer a sea change," come from a letter Mary Shelley sent to Maria Gisborne on May 2, 1823, the death of her husband having occurred a year earlier. The letter is reprinted in Paula R. Feldman and Diana Scott-Kilvert's *The Journals of Mary Shelley.*

E. J. Trelawny's first hearing of Percy Shelley while on Lake Geneva, as well as his subsequent meeting with Percy Shelley, Mary Shelley, and Lord Byron, with more extensive details, can be found in E. J. Trelawny's *Recollections of the Last Days of Shelley and Byron.*

The letter Percy Shelley sent E. J. Trelawny about the possibility of finding prussic acid is also reprinted in E. J. Trelawny's *Recollections of the Last Days of Shelley and Byron.* In this book are also Trelawny's memories of his friends' late departure from Leghorn, the mariner's dire warning, and the storms that ensued that day.

The search for Percy Shelley and Edward Williams and the eventual recovery of the two bodies are detailed in part in E. J. Trelawny's *Recollections of the Last Days of Shelley and Byron.* The cremation of the two bodies is also explained in gruesome detail in this book.

Captain Daniel Roberts's two letters to E. J. Trelawny upon the recovery of Shelley's boat, the *Don Juan,* are printed in E. J. Trelawny's *Recollections of the Last Days of Shelley and Byron.*

E. J. Trelawny's letter to Claire Clairmont following the death of Percy Shelley and his belief that Mary Shelley was partly to blame for the poet's unhappiness are copied in pencil in one of Captain Silsbee's memorandum books, in the Silsbee Papers.

Lord Byron's depression following Percy Shelley's death, his loss of weight, and his departure to Greece are detailed in part in Countess Blessington's *A Journal of Conversations with Lord Byron with a Sketch of the Life of the Author* and E. J. Trelawny's *Recollections of the Last Days of Shelley and Byron.* In this last book are also details of Byron's life on Missolonghi, the winter he spent there, his health decline, and his eventual passing, along with descriptions of the Greek doctors performing an autopsy on the poet and E. J. Trelawny's peek at the corpse before it was shipped off to England.

The search for the author of *Frankenstein* is described in the *Literary Panorama* and *British Critic.*

Descriptions and details of Mary Shelley's *The Last Man,* its writing, and its publication are to be found, in part, in Hugh J. Luke Jr.'s *The Last Man.*

EPILOGUE

Description of Alistair Cooke's death, his subsequent cremation, and the Cooke family's learning of the New York gang of body snatchers, along with details of the dismembering of his body, come from my own conversations with Susan

Cooke, Alistair Cooke's daughter, and also from conversations with Wendy Kogut, whose sister was a victim as well, and Patricia McNeill of the Brooklyn district attorney's office.

Details of Mary Shelley's trip to Italy in the company of her son Percy Florence and his friends come from Mary Shelley's *Rambles in Germany and Italy in 1840, 1842, and 1843.*

Details of Mary Shelley's death can be found in Betty T. Bennett's *The Letters of Mary Wollstonecraft Shelley;* Paula R. Feldman and Diana Scott-Kilvert's *The Journals of Mary Shelley;* Anne K. Mellor's *Mary Shelley: Her Life, Her Fictions, Her Monsters;* Muriel Spark's *Mary Shelley;* and Marion Kingston's *The Clairmont Correspondence: Letters of Claire Clairmont, Charles Clairmont, and Fanny Imlay Godwin.*

Eulogy of Mary Shelley published on February 15 of 1851 in *The Athenaeum.*

BIBLIOGRAPHY

Aldini, John. *An Account of the Late Improvements in Galvanism; with a Series of Curious and Interesting Experiments Performed Before the Commissioners of the French National Institute and Repeated Lately in the Anatomical Theatres of London; to Which Is Added, an Appendix, Containing the Author's Experiments on the Body of a Malefactor Executed at New Gate* (London: Cuthell & Martin, and J. Murray, 1803).

Aynsley, E. E., and W. A. Campbell. "Johann Konrad Dippel, 1673–1734," *Medical History* 6, no. 3 (1962): 281–86.

Bailey, James Blake, ed. *The Diary of a Resurrectionist, 1811–1812, to Which Was Added an Account of the Resurrection Men in London and a Short History of the Passing of the Anatomy Act* (London: Swan Sonnenschein & Co., 1896).

Ball, James Moores, MD, LLD. *The Body Snatchers: Doctors, Grave Robbers and the Law* (New York: Dorset House, 1989). Originally printed as *The Sack'em Up Men; An Account of the Rise and Fall of the Modern Resurrectionists* (Edinburgh and London: Oliver and Boyd, 1928).

Bennett, Betty T., ed. *The Letters of Mary Wollstonecraft Shelley*, vols. I–II (Baltimore: Johns Hopkins University Press, 1980, 1983).

Bernabeo, Raffaele, ed. *Luigi Galvani (1798–1998) fra Biologia e Medicina: Atti della Accademia delle Scienze dell'Istituto di Bologna, Classe di Scienze Fisiche, Anno 286* (Bologna: CLUEB, 1999).

Blessington, Countess. *A Journal of Conversations with Lord Byron. With a Sketch of the Life of the Author* (Boston: William Veazie, 1858).

Bloom, Harold. Afterword to a reprint of the 1831 edition of *Frankenstein; or, The Modern Prometheus* (New York: New American Library, 1965).

Boyle, Robert. *The Sceptical Chymist* (London: Everyman Series, 1661).

Bresadola, Marco. "Medicine and Science in the Life of Luigi Galvani (1737–1798)," *Brain Research Bulletin* 46, no. 5 (1998): 367–80.

———. "Animal Electricity at the End of the Eighteenth Century: The Many Facets of a Great Scientific Controversy," *Journal of the History of Neurosciences* 17, no. I (2008): 8–32.

Bresadola, Marco, and Giuliano Pancaldi, eds. *Luigi Galvani International Workshop Proceedings* (Bologna: University of Bologna International Centre for the History of Universities and Science, 1999).

Brockbank, William. "Old Anatomical Theatres and What Took Place Therein," *Medical History* 12 (1968): 371–84.

Byron, Lord George Gordon. *The Works of Lord Byron*, ed. Rowland Prothers (London: John Murray, 1904).

Cavallo, Tiberius. *A Complete Treatise on Electricity, in Theory and Practice, with Original Experiments, Containing the Practice of Medical Electricity, Besides Other Additions and Alterations* (London: Edward and Charles Diliy, 1777).

Crook, G. T., and John L. Rayner, eds. *The Complete Newgate Calendar* (London: Novarre Society, 1926).

Davy, Humphry. *A Discourse, Introductory to a Course of Lectures on Chemistry* (London: 1802).

———. *Collected Works*, ed. J. Davy, 9 vols. (London: Smith Elder, 1839).

Dickens, Charles. "A Visit to Newgate" (London: Chapman & Hall, 1835, 1877). Originally published in *Sketches by Boz*, 1835.

Dowden, Edward. *The Life of Percy Shelley*, 2 vols. (London: Paul Kegan, 1887).

Eriksson, Ruben, ed. *Andreas Vesalius First Public Anatomy at Bologna 1540. An Eyewitness Report* (Uppsala: Almquist & Wiksell, 1959).

Feldman, Paula R., and Diana Scott-Kilvert, eds. *The Journals of Mary Shelley, 1814–1844*, 2 vols. (Oxford: The Clarendon Press, 1987).

Ferrari, Giovanna. "Public Anatomy Lessons and the Carnival: the Anatomy Theatre of Bologna," *Past and Present* 117 (1987): 50–108.

Florescu, Radu. *In Search of Frankenstein* (London: Robinson Books, 2003).

Fox, John. *Vindication of Lady Byron* (London: Richard Bentley & Son, 1871).

Frati, Lodovico. *Il Settecento a Bologna. Presentazione di Mario Fanti* (Bologna: Atesa Editrice: 1978).

Fullner, June. *Young Humphry Davy: The Making of an Experimental Chemist* (Philadelphia: American Philosophical Society, 2000).

Galvani, Luigi. *De viribus electricitatus in motu musculari commentarius* (Bologna, 1791). Published in the seventh volume of the memoirs of the Institute of Sciences at Bologna.

Gelfand, Toby. "The 'Paris Manner' of Dissection: Student Anatomical Dissection in Early Eighteenth-Century Paris," *Bulletin of the History of Medicine* 46, no. 2 (1972): 99–130.

Glynn, R. Gryll. *Claire Clairmont—Mother of Byron's Allegra* (London: John Murray, 1939).

Godwin, William. *An Enquiry Concerning Political Justice* (London: G. G. J. and J. Robinson, 1793).

———. *Memoirs of the Author of a Vindication of the Rights of Woman*, second edition. (London: Printed for J. Johnson, No. 72, St. Paul's Church-yard, 1798).

———. "Correspondence & Papers, 1772–1836," Abinger Collection, Bodleian Library, Oxford University.

Grabo, Carl. *A Newton Among Poets—Shelley's Use of Science in "Prometheus Unbound"* (Chapel Hill: University of North Carolina Press, 1930).

Green, Robert M. "A Translation of Luigi Galvani's *De Viribus Electricitatus in Motu Muscularis Commentarius*. Commentary on the Effect of Electricity on Muscular Motion," *American Journal of the Medical Sciences* 227, no. 2 (1954): 231.

Grimm, Jacob. *Zeitung für einsiedler* (A Journal for Hermits), 1808.

Guerrini, Anita. "Anatomists and Entrepreneurs in Early Eighteenth-Century London," *Journal of the History of Medicine and Allied Sciences* 59, no. 2 (2004): 219–39.

Guiccioli, Teresa. *My Recollections of Lord Byron, and Those of Eye-Witnesses of His Life* (London: Richard Bentley, 1869).

Heilbron, J. L. "The Contributions of Bologna to Galvanism," *Historical Studies in the Physical and Biological Sciences* 2, no. 1 (1991): 56–85.

Hogg, Thomas Jefferson. *The Life of Percy Shelley* (London: Edward Moxon, 1858).

———. *Shelley at Oxford* (London: Mathuen & Co., 1904).

Holmes, Frederick L. "The Old Martyr of Science: The Frog in Experimental Physiology," *Journal of the History of Biology* 26, no. 2 (1993): 311–28.

Holmes, Richard. *Shelley: The Pursuit* (New York: E. P. Dutton, 1975).

Ireland, Thomas, Jr., ed. *The West Port Murders, or an Authentic Account of the Atrocious Murders Committed by Burke and His Associates; Containing a Full Account of All the Extraordinary Circumstances Connected with Them. Also, a Report of the Trial of Burke and M'Dougal with a Description of the Execution of Burke, His Confession, and Memoirs of His Accomplices, Including the Proceedings Against Hare & C.* (Edinburgh: Thomas Ireland Junior, 1829).

Jones, Frederick L., ed. *Maria Gisborne and Edward E. Williams: Shelley's Friends: Their Journals and Letters* (Norman: University of Oklahoma Press, 1951).

Kingston, Marion, ed. *The Journals of Claire Clairmont, 1814–1827* (Cambridge: Harvard University Press, 1968).

———. *The Clairmont Correspondence: Letters of Claire Clairmont, Charles Clairmont, and Fanny Imlay Godwin* (Baltimore and London: Johns Hopkins University Press, 1995).

Lawrence, Susan C. "Beyond the Grave: The Use and Meaning of Human Body Parts: A Historical Introduction." Published in *Stored Tissue Samples: Ethical, Legal, and Public Policy Implications*, ed. Robert F. Weir (Iowa City: University of Iowa Press, 1998).

Locke, Don. *A Fantasy of Reason: The Life and Thought of William Godwin* (London: Routledge & Keyon Paul, 1980).

Lucas, E. V., ed. *The Letters of Charles Lamb to Which Are Added Those of His Sister Mary Lamb* (London: J. M. Dent & Sons, 1935).

Lully, Raymond. *Experimenta*, reprinted in *Biographia Antiqua*, London, 1801.

MacDonald, D. L. *Poor Polidori: A Critical Biography* (Toronto: University of Toronto Press, 1991).

MacDonald, Helen. *Human Remains: Dissection and Its Histories* (New Haven, CT: Yale University Press, 2006).

Mackenzie, Peter. "The Case of Matthew Clydesdale." Printed in *Reminiscences of Glasgow and the West of Scotland*, vol. I (Glasgow: James P. Forrester, 1865).

Marchand, Leslie. *Byron: A Portrait* (Chicago: University of Chicago Press, 1979).

Medwin, Thomas. *The Life of Percy Bysshe Shelley*, ed. H. Buxton Formon (London: Oxford University Press, 1913).

Mellor, Anne K. *Mary Shelley: Her Life, Her Fiction, Her Monsters* (New York and London: Routledge, 1988).

Milton, John. *Paradise Lost*, book 10, 3rd Revised Edition (New York: W.W. Norton & Co., 2004).

O'Malley, C. D. *Andreas Vesalius of Brussels, 1514–1564* (Berkeley: University of California Press, 1964).

Orlandi, Pasquale. *Memorie Storiche Della Terra di Medicina* (Bologna: Atesa Editrice, 1852).

Otto, Peter. "The Regeneration of the Body: Sex, Religion and the Sublime in James Graham's Temple of Health and Hymen," *Romanticism on the Net* 23 (2001).

Paracelsus, *De Rerum Natura (Concerning the Nature of Things)*, in *The Hermetic and Alchemical Writings of Paracelsus the Great*, ed. and trans. Arthur Edward Waite (New Hyde Park, NY: University Books, 1967).

Pattison, F. L. M. "The Clydesdale Experiments: An Early Attempt at Resuscitation," *Scottish Medical Journal* (January 1986): 50–52.

———. *Granville Sharp Pattison: Anatomist and Antagonist, 1791–1851* (Tuscaloosa: University of Alabama Press, 2005).

Pera, Marcello. *The Ambiguous Frog* (Princeton, NJ: Princeton University Press, 1992).

Petrain, David. "An English Translation of John William Polidori's (1815) Medical Dissertation on *Oneirodynia* (Somnambulism)," *European Romantic Review* 21 (2010): 775–88.

Phillips, Charles. *Vacation Thoughts on Capital Punishments* (London: W. and F. G. Cash, 1757).

Porter, Roy. *English Society in the 18th Century*, rev. ed. (London: Penguin Books, 1991).

Quarterly Review, 18 (January 1818): 379–85.

Raffles, S. *Memoir of the Life and Public Services of Sir Thomas Stamford Raffles, F.R. S. & C., Particularly in the Government of Java 1811–1816, and of Bencoolen and Its Dependencies 1817–1824; with Details of the Commerce and Resources of the Eastern Archipelago, and Selections from His Correspondence* (London: John Murray, 1830).

Richardson, Ruth. *Death, Dissection and the Destitute* (Chicago: University of Chicago Press, 2001).

Rieger, James. "Dr. Polidori and the Genesis for Frankenstein," *Studies in English Literature, 1500–1900* 3, no. 4 (1963): 461–72.

Robinson, Charles E., ed. *The Frankenstein Notebooks: A Facsimile Edition* (Oxford: Routledge, 1996).

Robinson, Henry Crabb. *Diary, Reminiscences, and Correspondences of Henry Crabb Robinson, Barrister-at-Law, F.S.A.*, 2 vols., ed. Thomas Sadler, 3rd ed. (London and New York: Macmillan and Co., 1872).

Rossetti, Michael, ed. *The Diary of Dr. John William Polidori* (London: Elkin Mathews, 1911).

Scott, Sir Walter. *Blackwood's Edinburgh Magazine* 2 (20 March/April 1818): 613–20.

Shelley, Mary. *A Six Weeks' Tour Through a Part of France, Switzerland, Germany, and Holland, with Letters Descriptive of a Sail Round the Lake of Geneva, and of the Glaciers of Chamouni* (London: T. Hookham, Jr. and C. and J. Ollier, 1817).

———. *Frankenstein; or, The Modern Prometheus* (London: Lackington, Hughes, Harding, Mavor and James, 1818); repr. *Frankenstein or, The Modern Prometheus*, ed. James Rieger (Chicago: University of Chicago Press, 1982; New York: Bobbs-Merrill, 1974).

———. *Rambles in Germany and Italy in 1840, 1842, and 1843* (London: Edward Moxon, 1844).

———. *The Last Man*, ed. Hugh J. Luke Jr. (Lincoln: University of Nebraska Press, 1965).

———. *Frankenstein or, The Modern Prometheus*, ed. Maurice Hinoce (New York: Penguin Books, 1992).

Shelley, P. B. "Alastor," originally printed in 1816; reprinted in Raymond D. Havens, "Shelley's *Alastor*," *PMLA* 45, no. 4 (1930): 1098–1115.

Silsbee, Augustus Edward. Silsbee Papers, Memorandum Books, 1874–1875.

Simili, Raffaella. *Scienza a Due Voci* (Firenze: Leo S. Olschki, 2006).

Spark, Muriel. *Mary Shelley* (London: Constable, 1988).

Spedding, R. L. Ellis, and D. N. Heath, eds. *The Works of Francis Bacon*, 7 vols., 1879–90 edition (facsimile; Gale, *Making of Modern Law*, December 2010).

Trelawny, Edward John. *Recollections of the Last Days of Shelley and Byron* (Boston: Ticknor and Fields, 1859).

———. *Records of Shelley, Byron and the Author* (New York: Scribner and Welford, 1888).

Twain, Mark. *In Defense of Harriet Shelley and Other Essays* (New York and London: Harper & Brothers Publishers, 1918).

Ure, Andrew. "An Account of Some Experiments Made on the Body of a Criminal Immediately After Execution, with Physiological and Practical Observations," *Journal of Science and the Arts* 6 (1819): 283–94.

Vesalius, Andreas. *De Humani Corporis Fabrica Libri, Septum (On the Fabric of the Human Body)*, book I, "Bones and Cartilages," trans. William Frank Richardson in collaboration with John Berd Carman (Novato, CA: Norman Publishing, 2008).

Volta, Alessandro. *Elettricità Scritti Scelti* (Firenze: Giunti, 1999).

Walpole, Horace. "To Lady Ossory, Wednesday 23 August 1780," reprinted in *The Yale Editions of Horace Walpole's Correspondence*, ed. W. S. Lewis, vol. 33, p. II (New Haven, CT: Yale University Press, 1983).

Wilkinson, Charles Henry. *Elements of Galvanism in Theory and Practice*, 2 vols. (London: John Murray, 1804).

Wollstonecraft, Mary. *A Vindication of the Rights of Woman*, ed. Carl H. Poston, reprint edition (New York: W. W. Norton, 1975).

INDEX